W9-AEV-065

SPECTRA-STRUCTURE
CORRELATION

SPECTRA-STRUCTURE CORRELATION

by

JOHN P. PHILLIPS

DEPARTMENT OF CHEMISTRY
UNIVERSITY OF LOUISVILLE
LOUISVILLE, KENTUCKY

1964

ACADEMIC PRESS   New York and London

ACADEMIC PRESS INC.
111 Fifth Avenue, New York, New York 10003

United Kingdom Edition published by
ACADEMIC PRESS INC. (LONDON) LTD.
Berkeley Square House, London W.1

LIBRARY OF CONGRESS CATALOG CARD NUMBER: 64–20397

PRINTED IN THE UNITED STATES OF AMERICA

Preface

In view of the considerable number of comprehensive volumes on spectrophotometry already available, the prospective reader may justifiably wonder what prompted the production of this small volume.

The principal novelty of this book is its attempt to present a balanced survey of the data of absorption spectroscopy for organic compounds in all regions of the spectrum from far ultraviolet to far infrared. Classification of compounds according to functional groups is the basic indexing system employed.

A second thesis of this work is that spectra in all regions permit a uniqueness of identification of a compound that is not achieved with infrared or ultraviolet data alone. As a corollary of this point the more subtle, elaborate, and word-consuming analyses of data in each region may be replaced with brief statements citing the most striking effects in several regions. This accounts in part for the brevity of the treatment.

Most volumes on spectrophotometric subjects devote most of their space to instrumentation or theory, or occasionally both. Neither topic receives more than minimum comment here because I believe that instrumental problems should be turned over to instrument specialists and not to organic chemists and that theoretical discussions of the broad spectrum considered in this volume at even a minimum level would leave no room for the data.

This is a handbook of data and a survey of it. There has been unabashed use of secondary sources, to all of which I am deeply indebted.

JOHN P. PHILLIPS

Louisville, Kentucky
February, 1964

Contents

I

Introduction

II

Hydrocarbons

III

Compounds with Oxygen Function

IV

Compounds with Nitrogen Functions

V

Heterocycles

VI

Organic Compounds Containing Halogen, Sulfur, Phosphorus, Silicon, or Boron

VII

Inorganic Compounds

VIII

Complex Materials

I

Introduction

At a reasonable guess the number of compounds for which absorption spectra in some region have been published in the last twenty years is at least 200,000 and may be twice as many. Compilations of infrared spectra containing considerably more than 20,000 compounds and tables of ultraviolet absorption maxima for more than 50,000 compounds represent the current status of two of the continuing and successful efforts to collect current data.

The use of these vast numbers of spectra in any effective way for the characterization of molecular structure requires some sort of guiding principle. Here we have chosen to index the data according to the functional groups which are very familiar in organic chemistry, as far as this is possible.

A. Wavelength Classification of Absorption Spectra

Spectra from about 175 mμ in the far ultraviolet up to about 35 μ in the far infrared will be considered in the following chapters. Where data for all regions are available it is convenient to divide the discussion for each functional group or compound class into the following five categories, arranged in order of increasing wavelength of the absorbed radiation.

1. FAR ULTRAVIOLET

The 175 mμ lower limit is approximately the minimum wavelength obtained with a nitrogen-purged quartz spectrophotometer, and the 200 mμ upper limit is arbitrary. An occasional result from the shorter wavelengths attainable by vacuum spectrophotometers will be noted, but the ionization potentials computed from experimentation in the vacuum ultraviolet are not within the scope of the present work.

A significant part of the data below 195 mμ obtained with nitrogen-purged instruments not designed specifically for short wavelength operation is said to be invalidated by stray light effects,[1] and certainly different

[1] T. H. Applewhite and R. A. Micheli, *Tetrahedron Letters* **1961**, 560.

instruments have been shown to yield extraordinarily large variations in the spectrum shape of a standard compound.[2]

In view of the brevity of the wavelength range representing the far ultraviolet it is not surprising that the total number of published spectra in this region is a very few thousand.

2. ULTRAVIOLET–VISIBLE

Except for relatively minor instrumental changes there is no need to distinguish the ultraviolet from the visible, and the total wavelength range for both is arbitrarily chosen as 200–750 mμ. (An official upper limit of 780 mμ has been suggested.[3])

Absorption maxima in the visible can sometimes be estimated by eye (see Table I-1). The absence of one of the wavelength groups in the table

TABLE I-1
WAVELENGTH AND COLOR

Wavelength range (mμ)	Color of light	Complementary color
400–435 or 400–450[a]	Violet	Yellow-green
435–480 or 450–480	Blue	Yellow
480–490	Green-blue	Orange
490–500	Blue-green	Red
500–560	Green	Purple
560–580 or 560–575	Yellow-green	Violet
580–595 or 575–590	Yellow	Blue
595–610 or 590–625	Orange	Green-blue
610–750 or 625–750	Red	Blue-green

[a] Different publications list slightly different ranges since the color is partly subjective and dependent on the observer.

from the light transmitted by a solution normally gives the observer the impression of the complementary color, a solution strongly absorbing green light, for example, appearing purple. Since visual sensations are often complex mental interpretations of stimuli that are rarely pure transmitted light, the eye is not a reliable spectroscopic instrument. Solutions of compounds with more than one absorption band in the visible may have a misleadingly neutral appearance that changes greatly

[2] D. W. Turner *in* "Determination of Organic Structures by Physical Methods" (F. C. Nachod and W. D. Phillips, eds.). Vol. II, Chapter 5. Academic Press, New York, 1962.

[3] Editors, *Anal. Chem.* **33**, 1968 (1961).

with illumination, concentration, thickness and other factors. Thus, aqueous solutions of chromic salts that are blue-green by fluorescent light may be red under tungsten or other types of light source. Differences of color produced by reflected rather than transmitted light are common.

3. NEAR INFRARED

The range is 0.75 μ (750 mμ) to 3 μ, though certainly the lower end overlaps the performance capabilities of most visible region spectrophotometers and the upper end from 2 to 3 μ is within the range of standard infrared instruments. However, the resolution of the salt optics used in many infrared spectrophotometers is inferior to that of the quartz prism or grating systems employed in near infrared instruments, and the latter provide better data. Perhaps it would be more nearly accurate to define the near infrared as the 0.75–3 μ range just because this part of the spectrum is conveniently measured with a photoconductive cell detector.

4. INFRARED

The salt optics region is approximately 2–15 μ (5000–666 cm^{-1}) with an occasional extension as high as 16–18 μ. Sodium chloride prisms give best results at the long end of this range of wavelengths, but the most useful part for functional group correlations is the short end. Prisms of lithium fluoride or the use of grating instruments give superior results at short wavelengths, and the present trend of commerically available models is to gratings with a vastly better performance than ever offered before.

5. FAR INFRARED

This will generally be the 15–35 μ range available to cesium bromide optics, though there are significant data to much longer wavelengths that will occasionally be cited.

The most appropriate unit for wavelength is a troublesome choice, and from the viewpoint of significant figures alone cannot be the same in all parts of the available spectrum. A decision between frequency units, which are proportional to energy, and wavelength units which are not is also required.

In the following chapters we will use the millimicron, mμ, as the unit of wavelength for the far ultraviolet, ultraviolet, and visible and avoid the wavenumber (kayser) frequency unit even though the latter is preferred by modern inorganic and coordination chemistry. For the near infrared standard practice appears to favor the micron, μ, but in the ordinary infrared the literature shows an overwhelming preference for the wavenumber frequency unit, cm^{-1}. As a result of this difference the reader may suffer some annoyance in sections of following chapters where near infrared and infrared discussions overlap. In the far infrared the micron appears to be narrowly favored over the wavenumber.

The following relationships of units may be helpful:

$$1 \ \mu \ = \ 1000 \ \text{m}\mu \ = \ 10{,}000 \ \text{Å} \ = \ 10^{-4} \ \text{cm} \ = \ 10{,}000 \ \text{cm}^{-1}$$

B. Intensities and Shapes of Absorption Bands

The molar absorptivity, ϵ, is the nearly universal measure of light absorption for compounds of known molecular weight and is defined in terms of the Beer-Lambert law:

$$A \ = \ \epsilon b c$$

where A is the measured absorbance, b the path length of the sample in centimeters and c its molar concentration. The units for ϵ are then liters per mole per centimeter.

In general the largest values of molar absorptivity are found in far ultraviolet and ultraviolet spectra, and successful experimental work may therefore require very dilute solutions. Bands in near and far infrared spectra are commonly very feeble, and high concentrations or even the pure compounds at considerable thickness are consequently required.

The range of molar absorptivities for electronic spectra (far ultraviolet, ultraviolet–visible) is roughly 0.001 to more than 400,000, with the lower limit experimentally controlled by solubility or impurity factors and the upper limit theoretically controlled by the size of the absorbing unit and the probability of the electronic transition (see below). In the near infrared molar absorptivities of 1–10 are fairly large and values much above a hundred rare; since most of these bands are overtones and shorter wavelengths represent higher and weaker overtones, it is not surprising that the higher absorptivities usually fall at the longer wavelengths. Infrared molar absorptivities are rather infrequently calculated, but values in the hundreds are rated strong.

The molar absorptivity at a wavelength of maximum absorption may vary considerably with solvent, temperature, phase, and other factors, but the area under the whole band is often much more nearly constant and thus may be a better measure of true absorptivity. This area is the integral of $\epsilon d\nu$ where ν is the frequency. A related factor of considerable theoretical interest is the oscillator strength, f, a measure of the number of oscillators producing the absorption band; for smooth and symmetrical absorption bands in electronic spectra f may be computed from the band width, $\Delta\nu$, at half-maximum molar absorptivity by the equation[4]:

$$f = 2.2 \times 10^{-9} \, \Delta\nu\epsilon_{max} \qquad (\Delta\nu \text{ in cm}^{-1})$$

Although the band area has theoretical significance and the molar absorptivity at a maximum does not, the latter is much easier to determine (though with a random error of perhaps 10% in the most unfavorable region, infrared).

The shape of an absorption band may be informative on occasion. For example, the measured width of the band in electronic spectra can indicate the degree of strain or distortion in a soluble complex,[5] and certainly the shape serves for the detection of overlapping or partially submerged absorptions in all regions. The numerous journals of the American Chemical Society have long banned graphs of spectral data when offered for characterization purposes only, a policy that no doubt saves space but tends to inhibit attempted analyses of band shapes.

C. Instruments

The history of spectrophotometry begins well before 1900, and rather large collections of infrared spectra by Coblentz and of ultraviolet by Jones and others before 1910 are not only pioneering ventures but very respectable work by any standard.[6] However, in the last twenty years spectrophotometry has changed its character from a research project in experimental physics to a routine technical service in many laboratories, a change due largely to the existence of commercial photoelectric spectrophotometers.

Since this book aims only at the analysis of spectrophotometric data, an indication of the approximate range, precision, and accuracy of standard commercial instruments is the only attention to instrumentation that seems justifiable.

Single instruments that can cover far ultraviolet, ultraviolet, visible and near infrared regions with only minor modifications and single

[4] S. F. Mason, *Quart. Rev. (London)* **15**, 287 (1961).
[5] S. M. Crawford, *Spectrochim. Acta* **18**, 965 (1962).
[6] R. A. Morton, *J. Roy. Inst. Chem.* **84**, 5 (1960).

models for both infrared and far infrared are commercially available at present, and it is altogether probable that one instrument for the full range from far ultraviolet to far infrared will soon be designed.

In Table I-2 is listed a representative group of ultraviolet-visible spectrophotometers; it is evident from the wavelength ranges given that the Beckman DK's, Cary 14, Perkin-Elmer 350 and Zeiss PMQ II are the so-called "all-region" models for service from far ultraviolet through near infrared. The Beckman B and DU (both rather old models) are manual, but the others have automatic recording systems.

TABLE I-2

SPECTROPHOTOMETERS FOR ULTRAVIOLET, VISIBLE, AND NEAR INFRARED

Model	Range (mμ)	Accuracy	
		Wavelength (Å)	Photometric[a]
Bausch & Lomb Spectronic 505	200–700	5	0.005
Beckman B	320–1000	5–20[b]	0.5% T
Beckman DB	220–700	10–100	1% T
Beckman DK-2	200–3000	1[b] (uv)	0.01
DK-2A	170–3500		
Beckman DU	200–1000	0.5–5[b]	0.003[c]
Cary Model 11	200–800	<5–<10	0.004
Cary Model 14	186–2600	<4	0.002
Cary Model 15	185–800	1–10	0.002–0.005
General Electric	380–700	<10	0.5%
Perkin-Elmer 202	190–750	5–10	0.01
Perkin-Elmer 350	175–2700	1–5 (uv–visible)	—
Zeiss PMQ II	200–2500	0.5 (at 250 mμ)	—

[a] In units of absorbance except as shown.

[b] Resolution; accuracy is about half as much. In prism instruments where a range is given the lower figure is for the shorter wavelengths.

[c] L. Cahn, *J. Opt. Soc. Am.* **45**, 953 (1955).

For the infrared and far infrared there are many double beam recording spectrophotometers (Table I-3) of varying price levels and types. Many feature salt optical systems with interchangeable prisms to allow performance in more than one segment of the infrared. Combinations of prisms with gratings and of gratings with filters have been used in recent years to obtain instruments of high resolution. Accuracy to the nearest wavenumber is available in some of the newer grating models, a performance level far better than any commercial instrument could offer even five years ago, and modifications for even finer work have been successfully tried.[7] Improvements in both ultraviolet and infrared instruments are exceedingly rapid, and these tables may well be obsolete practically at the date of publication.

TABLE I-3

REPRESENTATIVE INFRARED SPECTROPHOTOMETERS

Instrument	Range $(\mu)^a$	Accuracy	Type
Baird NK-1	2–16+	$0.015\ \mu$; $0.5\%\ T$	Prism
Baird NK-3	2–22	1–4 cm^{-1}; $0.5\%\ T$	Prism-grating
Beckman IR 4	1–15+	$0.015\ \mu$; $1\%\ T$	Prism
Beckman IR 5A	2–16	$0.025\ \mu$; $1\%\ T$	Prism
	11–35	$0.04\ \mu$	CsBr prism
Beckman IR 7	0.7–15.5+	0.5–5 cm^{-1}	Grating
Beckman IR 8	2.5–16	0.008–0.015 μ	Filter-grating
Beckman IR 9	2.5–25+	0.2–0.6 cm^{-1}	Prism-grating
Cary 90	2.5–22	0.5 cm^{-1}	Prism-grating
Perkin-Elmer 21	2.5–15.5	1.5–20 cm^{-1}	Prism
Perkin-Elmer 137-B	2.5–15	$0.03\ \mu$; $0.5\%\ T$	Prism
Perkin-Elmer 137-G	0.83–7.65	0.004–0.012 μ	Grating
Perkin-Elmer 221	0.7–38		Prism-grating
Perkin-Elmer 301	15–167		Grating
Perkin-Elmer 421	5–40	1 cm^{-1}	Filter-grating

^{a} The plus sign indicates possible range extension with an auxiliary prism.

D. Mechanisms for the Absorption of Radiation

An extensive discussion of the elaborate theoretical framework of spectrophotometry will not be attempted. Simple molecules containing only a few atoms have spectra that can be very nearly completely accounted for by suitable theories, and volumes have been written about them.[8] For more elaborate molecules a full theoretical treatment is much less likely to be available, and here only a few easy generalizations will be considered.

According to elementary quantum concepts the energy of electromagnetic radiation increases linearly with frequency. The frequencies of the infrared part of the spectrum are short and thus interact only feebly

[7] P. J. Krueger, *Appl. Opt.* **1**, 443 (1962).

[8] G. M. Barrow, "Introduction to Molecular Spectroscopy." McGraw-Hill, New York, 1962.

with absorbing matter, effecting vibrations and rotations of molecules and submolecular groups. Rotation requires less energy than vibration and is observed in a relatively pure form in the far infrared spectra of simple compounds, but most of the ordinary infrared spectrum is associated with vibrational effects with superimposed rotational components. The near infrared is an unusual part of the spectrum in that harmonics, overtones and combinations of fundamental infrared frequencies for functional groups containing hydrogen are its most significant components (though the lower wavelength region also contains some low energy electronic transitions overlapping from the visible).

In the far ultraviolet, ultraviolet, and visible regions of the spectrum the predominant effect of absorbed radiation, because of its greater energy as compared to the infrared, is the excitation of electrons in molecules of sufficiently complex structure to have one or more such transitions needing a moderate amount of energy. Transitions requiring the least energy produce absorption bands at the longest wavelengths, sometimes reaching the lower portion of the near infrared. Vibrational components may sometimes be observed in electronic spectra in the form of fine structure, a series of closely spaced maxima that may number in the hundreds for some compounds in the vapor phase when measured with a high resolution spectrophotometer.

Compounds with absorption bands in the visible almost invariably have other bands in the ultraviolet as well, but the number of substances with ultraviolet maxima only is many times as large.

1. Electronic Spectra

The most common electronic transitions are indicated in the following chart (Table I-4) of electron symbols, energies, and effects. This classification, though rudimentary, is very helpful in the interpretation of ultraviolet–visible bands, as is evident throughout this volume.

Most of the $\sigma \to \sigma^*$ transitions require such high energy that they are found only below 200 mμ, in the far ultraviolet; an example is the absorption band in the vacuum ultraviolet shown by saturated aliphatic hydrocarbons. The most familiar examples of $\pi \to \pi^*$ spectra are the conjugated polyenes in which the transition energy declines with increasing length of the conjugated system and the wavelength of maximum absorption correspondingly increases. The $\pi \to \pi^*$ bands are usually strong and obviously require an unsaturated molecular structure. The $n \to \pi^*$ effects require less energy and thus appear at longer wavelengths; the weak bands of carbonyl, nitro, nitroso, and similar compounds containing both multiple bonds and nonbonding electrons on at least one atom comprising a multiple bond are very distinctive, and some of these bands are in the visible spectrum. The $n \rightarrow \sigma^*$ transition requires enough energy to appear usually at the shorter wavelengths of the ultraviolet; an example is the absorption of an alkyl halide for which nonbonding electrons are supplied by the halogen, and the maximum increases in wavelength in the order Cl < Br < I as the electrons are successively easier to excite.

TABLE I-4

ELECTRONIC TRANSITIONS AND SPECTRA

Electron level[a]	Symbol	Transitions	Remarks
Antibonding sigma	σ^*	$\sigma \rightarrow \sigma^*$	Far ultraviolet; only transition for saturated molecules without n electrons.
		$n \rightarrow \sigma^*$	Usually at short wavelengths of the ultraviolet.
Antibonding pi	π^*	$\pi \rightarrow \pi^*$	Prominent in conjugated systems ("K-bands"); often very strong.
		$n \rightarrow \pi^*$	At relatively long wavelengths and usually weak.
Nonbonding	n		
Bonding pi	π		
Bonding sigma	σ		

[a] In order of decreasing energy from top to bottom.

There are several systems of nomenclature for electronic transitions at present. The designations "K-band" for a conjugated system pi-electron transition and "R-band" for the $n \rightarrow \pi^*$ transition are considered rather out of date. The terms $N \rightarrow V$ (sometimes the arrow is reversed) for transitions of sigma and pi electrons and $N \rightarrow Q$ for those of nonbonding electrons are often employed. Many now favor the precise designation of electron levels by letters with both superscript and subscript notations, especially for aromatic systems (see Chapter II).

The molar absorptivity of a compound is a function of the cross-sectional area of the absorbing species and the probability of the necessary electron transition. For an average organic molecule with an assumed cross-section of about 10^{-15} cm^2 a unit transition probability has been calculated[9] to correspond to a molar absorptivity of the order of

[9] E. A. Braude, *in* "Determination of Organic Structures by Physical Methods" (E. A. Braude and F. C. Nachod, eds.). Academic Press, New York, 1955.

10^5, and the highest known values are indeed a few hundred thousand.[10] Any molar absorptivity above 10,000 is conventionally considered high and a value under 1000 low.

2. INFRARED SPECTRA

Frequencies for vibrations of individual functional groups are mostly above 1000 cm^{-1} and over-all molecular vibrations usually below this value. A summary of functional group frequencies is given in Table I-5.

The stretching frequencies for functional groups composed of two atoms, A and B, may be roughly estimated by assuming that the bond between them is a spring vibrating in accord with Hooke's law. The following equation may be used to calculate the frequency:

$$\nu(cm^{-1}) = 1307(k/\mu)^{1/2}$$

with ν the wavenumber of the band, k a force constant factor usually given the values 5, 10, and 15, respectively, for single, double, or triple bond, and μ is the reduced mass of the A-B pair in atomic weight units, $m_A m_B/(m_A + m_B)$. If A or B is hydrogen, disputed or difficult frequency correlations may sometimes be checked by substituting deuterium for hydrogen. Since deuterium is twice as heavy as hydrogen a substantial displacement of stretching frequency is calculated by the above equation; for example, the C—H stretching values are near 3000 cm^{-1} and the C—D values near 2100 cm^{-1}.

The force constant of the A—B bond is increased by substituents that tend to increase the number of electrons participating in the bond, and the result is a slight increase in the corresponding stretching frequency. Effects of this nature account for the characteristic differences of the carbonyl band position in various classes of aldehydes, ketones, and related compounds. Correlations of such frequency shifts with Hammett sigma constants,[11] group electronegativities,[12] ionization potentials,[13] chemical shifts in NMR spectra or other measures of bond polarity or electron distribution can be made with some success. Indeed, the extensive application of these correlations is one of the more important trends in recent efforts to determine structure from spectra.

[10] J. B. Armitage, E. R. H. Jones, and M. C. Whiting, *J. Chem. Soc.* **1952**, 2014; F. Sondheimer, R. Wolovsky, and Y. Amiel, *J. Am. Chem. Soc.* **84**, 274 (1962).

[11] H. H. Freedman, *J. Am. Chem. Soc.* **82**, 2454 (1960); M. St. C. Flett, *Trans. Faraday Soc.* **44**, 767 (1958); M. J. S. Dewar and P. J. Grisdale, *J. Am. Chem. Soc.* **84**, 3546 (1962).

[12] J. K. Wilmshurst, *Can. J. Chem.* **38**, 467 (1960).

[13] D. Cook, *Can. J. Chem.* **39**, 31 (1961).

In hydrogen bonded structures the functional group carrying the hydrogen must hold it more weakly than normal; thus, the force constant is effectively diminished and the absorption band is at lower frequency than for the same group with nonbonded hydrogen. Since hydrogen bonding may be intra- or intermolecular in some compounds and a portion of the hydrogens may remain unbonded, the structure of the O—H stretching region for phenols and alcohols (for example) is sometimes very informative.

When two identical A—B bonds have one atom in common, e.g.,

TABLE I-5

APPROXIMATE FREQUENCIES OF FUNCTIONAL GROUPS IN THE INFRARED

Frequency (cm^{-1})	Type of group[a]	Borderline groups
2000–5000	X—H stretching X≡X stretching Z—H stretching X=X=X stretching X=X=Y stretching	
1400–2000	X=X stretching (aliphatic) X=X stretching (aromatic) X=Y stretching	
		X—H bending X—H wagging
1000–1400	X—X stretching X—Y stretching X=Z stretching Z—H bending	
		X—X stretching X—Y cyclic
650–1000	X=X—H bending XH$_2$ rocking X—H aromatic bending X—Z stretching	
400–650	X—Z bending X—X and X—Y bending (in cyclic or aromatic compounds) X—Br stretching X—I stretching	

[a] X and Y may be carbon, nitrogen, oxygen, or fluorine; Z may be silicon, phosphorus, sulfur, or chlorine. Entries are in order of decreasing frequency.

CF_2, CH_2, C—O—C, NH_2, SO_2, the characteristic vibrational frequency is usually split into two bands quite close to each other, one representing the symmetric and the other the asymmetric vibration of the group. For stretching frequencies the asymmetric is usually higher, but for bending modes the symmetric is higher.

Compounds of high molecular symmetry are likely to have only a few infrared bands because a vibration or rotation that does not produce a dipole moment change is not infrared active. The relative transparencies of carbon tetrachloride and disulfide in most of the infrared are explained in this way.

The stretching frequencies of individual functional groups are generally above 1000 cm^{-1}, while bending and deformation processes, which are likely to be less helpful for structure analysis, are often at shorter frequencies (Table I-5). However, it is the short frequency region that has the well-known fingerprint characteristic; it is generally believed that no two compounds are identical in spectrum in this region, though neighboring members at the high end of an homologous series may call for a sharp eye to detect the differences.

Designations of infrared bands as strong, medium, or weak are usually made by internal comparison, with the highest maximum labeled strong even though its actual molar absorptivity may be rather low. Obviously comparisons of different spectra on this basis can be very misleading, but it has been quite difficult to determine satisfactory molar absorptivities in the infrared.

a. Near infrared.

Overtones of the fundamental stretching frequencies of C—H, O—H, or N—H bonds together with various combination bands are the main components of near infrared spectra. The fundamentals for all three bonds lie near 3 μ, and the overtones are readily calculated as successive integral multiples of the fundamental frequencies. As the integral multiplier increases, the intensity of the corresponding overtone generally decreases considerably, so that only the first couple of overtones are likely to have appreciable absorption. The O—H and N—H bands are usually considerably stronger than the C—H values. In the 2.5–3.0 μ region molar absorptivities as high as several hundred may be found; from 1.4 to 2.5 μ the usual values are 0.1–3.0; and below 1.4 μ values of only a few hundredths of a unit are common.

Often the strongest bands in a near infrared spectrum come from combinations, either the sum or difference of two fundamental frequencies or the sum or difference of a fundamental and an overtone. Although it is often difficult to ascertain the source of a given combination band, the following example is suggestive. For the —CH$_2$— group the stretching fundamental (for C—H) lies near 2900 cm^{-1} and the symmetrical deformation near 1460 cm^{-1}; the sum of these frequencies accounts for a 2.3 μ band in the near infrared; similarly the sum of the stretching fundamental and the first overtone of the deformation frequency gives a band near 1.73 μ, and the sum of the deformation frequency and the first overtone of the stretching fundamental accounts for a 1.39 μ band, and so on. These combinations are observed in aliphatic hydrocarbons and their derivatives.

The lower wavelength end of the near infrared contains bands with easy electronic transitions as source; for example, many coordination compounds of copper and cobalt absorb here.

b. Far infrared.

Very simple molecules may yield pure rotational spectra in the far infrared, but for the usual organic compound various skeletal bendings are probably the most significant source of absorption. These may be expected to have a high degree of uniqueness from compound to compound, and few broad generalizations are possible in this part of the spectrum.

E. Sources of Spectrophotometric Data

The total number of spectra in the published literature is hard to estimate because the data are so widely scattered and poorly dealt with by Chemical Abstracts. Considerably more than 20,000 ultraviolet–visible spectra were published in 1958–1959 (according to the collection made by Organic Electronic Spectral Data, Inc. for those years), about 9000 in the Journal of the American Chemical Society, the Journal of Organic Chemistry and the Journal of the Chemical Society combined.

For the period 1946–61 the files of Organic Electronic Spectral Data, Inc., permit an estimate of 100,000–200,000 ultraviolet–visible spectra; probably the number of infrared spectra is at least comparable.

1. Ultraviolet–Visible Spectra

In addition to a number of older collections of data such as those in International Critical Tables and other encyclopedic publications, the following relatively modern compilations are noteworthy either for size or comprehensive coverage of some particular field. Advances in instrumentation have been so rapid that these modern collections should always be preferred.

(a) "Organic Electronic Spectral Data," Vol. I (M. J. Kamlet, ed.) Vol. II (H. E. Ungnade, ed.). Interscience, New York, 1960. Tables of λ_{max} and corresponding log ϵ collected from 60 journals published 1946–1955. Data for about 50,000 compounds are listed in empirical formula order. Volumes III and IV cover the literature up to 1960, adding perhaps 40,000 compounds, and further volumes are in preparation. This is the largest single collection of spectrophotometric data of any kind.

(b) "Sadtler Ultraviolet Spectra," sold by Samuel P. Sadtler and Son, Inc., Philadelphia, Pa. This collection was started on a subscription basis in 1959 as an adjunct to the larger infrared collection and currently provides reproductions of a few thousand spectra. The price and the subscription are high in relation to book prices.

(c) American Society for Testing Materials (ASTM). Punched cards indexing ultraviolet bands to the nearest 2 mμ. There are about 10,000 cards in the deck at present and periodic additions are made. An IBM sorter is required for effective use of the cards and the price is rather high.

(d) "Catalog of Ultraviolet Spectral Data," American Petroleum Institute Project 44, Agricultural and Mechanical College of Texas, College Station, Texas. At the end of 1962 this collection totaled 870 loose-leaf reproductions of spectra, mainly of hydrocarbons and other compounds of interest in petroleum chemistry. The Manufacturing Chemists' Association Research Project (same address) has begun a similar collection of more general spectra, but only a few have been published so far. Both collections are distributed free to appropriate organizations.

(e) L. Lang, ed., "Absorption Spectra in the Ultraviolet and Visible Region," 2 vols. Academic Press, New York, 1961. About 400 spectra of important compounds are given in detail. Additional volumes in this series can be expected.

(f) A. Eucken and K. H. Hallwege, eds., "Landolt-Börnstein Zahlenwerte und Funktionen aus Physik, Chemie, Astronomie, Geophysik und Technik," 6th. ed. In part three of the first volume of this well known reference work the article on ultraviolet–visible spectra contains over a thousand representative spectra as well as 2639 classified references to others.

(g) R. A. Friedel and M. Orchin, "Ultraviolet Spectra of Aromatic Compounds." Wiley, New York, 1951. An atlas of 579 uniformly graphed spectra of aromatic compounds of various types.

(h) H. M. Hershenson, "Ultraviolet and Visible Absorption Spectra, Index for 1930–1954"; "Index for 1955–1959." Academic Press, New York, 1956; 1961. No spectra are reproduced or listed, but references to the original literature are indexed by compound name. Over 32,000 references are cited in the first of these two volumes alone.

A number of volumes or long review articles on the interpretation of ultraviolet–visible spectra are so heavily documented that they may be considered useful data collections as well. Only the more recent and valuable of these are listed below.

(a) H. H. Jaffé and M. Orchin, "Theory and Applications of Ultraviolet Spectroscopy." Wiley, New York, 1962. By any standard this is the largest and most nearly comprehensive review of the subject yet published.

(b) A. E. Gillam and E. S. Stern, "Introduction to Electronic Absorption Spectroscopy in Organic Chemistry," 2nd ed. Edward Arnold, London, 1957. There are 163 tables and 94 graphs as well as a systematic discussion of spectra-structure correlation in this standard work. It was published too early to have the use of the large data compilations mentioned above, and the approach is more empirical than the more modern books.

(c) C. N. R. Rao, "Ultraviolet and Visible Spectroscopy." Butterworths, London, 1961. This resembles Gillam and Stern but is shorter and more up to date.

(d) S. F. Mason, *Quart. Rev.* **45**, 287–371 (1961). This lengthy review gives meticulous attention to the nature of each electronic transition that yields an absorption band.

(e) G. H. Beaven, E. A. Johnson, H. A. Willis, and R. G. J. Miller, "Molecular Spectroscopy." Macmillan, New York, 1961. The first half (by the first two authors) summarizes all aspects of ultraviolet spectrophotometry. The bibliography includes a great many of the older collections that are not described here. The second half of the volume is about infrared spectra.

(f) C. K. Jørgensen, "Absorption Spectra and Chemical Bonding in Complexes." Pergamon, Oxford, 1962. Though only inorganic and coordination compounds are within the scope of this volume there are numerous tables of absorption bands (with kilokayser frequency units) with assignments for the electron transitions. The book is a striking illustration of recent progress in understanding the spectra of transition metal compounds.

(g) E. A. Braude and F. C. Nachod, eds., "Determination of Organic Structures by Physical Methods." Academic Press, New York, 1955. See the chapter by E. A. Braude.

Reviews dealing with special topics and generally having very extensive bibliographies for the spectra discussed include the following, listed by subject.

(a) Amino acids. *In* "Chemistry of the Amino Acids" (J. P. Greenstein and M. Winitz, eds.), pp. 1688–1722. Wiley, New York, 1961.

(b) Aromatic hydrocarbons. Numerous papers and a book (see Chapter II) by E. Clar. Also R. N. Jones, *Chem. Revs.* **32**, 1–46 (1943).

(c) Charge transfer spectra. L. Orgel, *Quart. Rev. (London)* **8**, 422 (1954).

(d) Nucleic acids and related materials. G. H. Beaven, E. R. Holiday, and E. A. Johnson, *in* "The Nucleic Acids" (E. Chargaff and J. N. Davidson, eds.), Vol. I. Academic Press, New York, 1955.

(e) Polymers. M. Tryon and E. Horowitz *in* "Analytical Chemistry of Polymers," Part II (G. M. Kline, ed.). Interscience, New York, 1962.

(f) Proteins. G. H. Beaven, *Advan. Spectry.* **2**, 331 (1961).

(g) Stereochemistry. E. A. Braude and E. S. Waight, *Progr. Stereochem.* **1**, 126 (1954).

(h) Steroids. L. Dorfman, *Chem. Revs.* **53**, 47–144 (1953).

2. INFRARED SPECTRA

Collections of actual spectra rather than tables of data are preferred because the average infrared spectrum has so many maxima.

The ASTM punched card deck of infrared spectra has indexed data from most of the important collections, including the Sadtler collection of well over 20,000 compounds (itself indexed by name, formula, function, and strongest maximum and with machine search facilities), the over 2400 loose leaf spectra (at the end of 1962) of the American Petroleum Institute Project, the Coblentz society collection, and the relatively small N.B.S. and D.M.S. (Documentation of Molecular Spectroscopy) groups. Other large compilations from the older literature have been summarized in part by Jones and Sandorfy,[14] and special atlases and collections for steroids, polymers and other special materials will be noted in appropriate later sections of this book.

The interpretation of infrared absorption bands has been considered

[14] R. N. Jones and C. Sandorfy *in* "Chemical Applications of Spectroscopy" (W. West, ed.). Interscience, New York, 1956.

at some length in nearly every volume or series of volumes concerned with physical methods (e.g., those edited by Gilman, Berl, Weissberger, Houben-Weyl, etc.), but the standard work for simple empirical correlation has been L. J. Bellamy's "The Infrared Spectra of Complex Molecules," 2nd ed. Wiley, New York, 1958. An equally exhaustive treatment is the very long chapter by R. Norman Jones and C. Sandorfy, *in* "Chemical Applications of Spectroscopy" (W. West, ed.). Interscience, New York, 1956 (Volume IX of the Weissberger series); as an appendix it lists group frequencies in decreasing numerical order to facilitate identification of functional groups in unknown compounds. A brief set of correlation charts and tables with minimum discussion has been provided by A. D. Cross in "Introduction to Practical Infrared Spectroscopy." Butterworths, London, 1960.

The most interesting device for summarizing functional group absorption bands is the Colthup chart,[15] a plot of functional group frequency ranges as estimated from average results for a large number of compounds containing the function. The original chart has been reproduced in nearly every instrumental text published in the last ten years. If a chart of this type is drawn to the same scale as actual spectra recorded in the laboratory and is reasonably transparent, it may be superimposed on the spectra to facilitate comparisons. Beckman Instruments and the Dow Chemical Company among others have distributed large charts that can be used in this way.

An exhaustive list of Colthup-type chart presentations has not been compiled, but the following list indicates the subjects of a number of rather recent examples. Several also appear in later chapters of this book.

Groups of several charts.	L. J. Bellamy, "The Infrared Spectra of Complex Molecules," 2nd ed. Wiley, New York, 1958. (5 charts).
	A. D. Cross, "Introduction to Practical Infrared Spectroscopy." Butterworths, London, 1960. (6 charts).
	L. E. Kuentzel, from C. N. Reilly and D. T. Sawyer, "Experiments for Instrumental Methods." McGraw-Hill, New York, 1961.
Near infrared.	R. F. Goddu and D. A. Delker, *Anal. Chem.* **32**, 140 (1960).

---

[15] N. B. Colthup, *J. Opt. Soc. Am.* **40**, 397 (1950).

Far infrared of hydrocarbons.	F. F. Bentley and E. F. Wolfarth, *Spectrochim. Acta* **15**, 165 (1959).
C—H stretching frequencies.	S. Wiberley, S. C. Bunce, and W. H. Bauer, *Anal. Chem.* **32**, 217 (1960).
C—O—C vibrations, 1000–1250 cm^{-1}.	H. Hoyer *in* "Methoden der Organischen Chemie (Houben-Weyl)" (E. Müller, ed.), Vol. III, Part 2, p. 850.
Alkyl phenols.	D. D. Shrewsbury, *Spectrochim. Acta* **16**, 1298 (1960).
Thiocarbonyl compounds.	C. N. R. Rao and R. Venkataraghavan, *Spectrochim. Acta* **18**, 541 (1962).
Phosphorus compounds.	D. E. C. Corbridge, *J. Appl. Chem.* **6**, 456 (1956); *J. Chem. Soc.* **1954**, 494, 4558.
Organosilicon compounds.	A. L. Smith, *Spectrochim. Acta* **16**, 104 (1960); see also R. N. Kniseley, V. A. Fassel, and E. E. Conrad, *ibid.* **15**, 652 (1959).
Polymers.	M. Tryon and E. Horowitz *in* "Analytical Chemistry of Polymers, Part II." (G. M. Kline, ed.). Interscience, New York, 1962.
Inorganic compounds.	S. R. Yoganarasimhan and C. N. R. Rao, *Chemist-Analyst* **51**, 21 (1962); J. Ferraro, *J. Chem. Educ.* **38**, 206 (1961).
Inorganic compounds (far infrared).	F. A. Miller, G. L. Carlson, F. F. Bentley, and W. H. Jones, *Spectrochim. Acta* **16**, 162 (1960).

If these charts are adequately spaced, there may be room to note band peculiarities or relative intensities; the latter may also be indicated by the thickness of the line denoting the frequency range. For individual compounds schematic spectra in which a vertical line with height proportional to intensity takes the place of each absorption band have the advantage of taking less space than the actual spectra and of offering more specific information than the Colthup chart.

Hershenson has compiled indexes[16] of infrared spectra from the literature for 1945–1957 that resemble his corresponding ultraviolet indexes already discussed. They are perhaps more valuable than the ultraviolet indexes because it appears unlikely that any organization will ever tackle the formidable task of listing infrared maxima for many thousands of compounds as Organic Electronic Spectral Data, Inc., is doing for the

[16] H. M. Hershenson, "Infrared Absorption Spectra, Index for 1945–1957." Academic Press, New York, 1959.

ultraviolet. There is also a British government publication, "An Index of Published Infra-Red Spectra," in several volumes.

3. FAR ULTRAVIOLET AND NEAR AND FAR INFRARED

The far ultraviolet has been reviewed by D. W. Turner,[2] with full attention to the pioneer reviews of Platt and Klevens.

The near infrared has been reviewed by Goddu,[17] Wheeler[18] and Kaye.[19] Goddu is far the most useful for correlation studies, but Wheeler cites references to several hundred compound spectra listed in empirical formula order.

A bibliography[20] of far infrared articles up to 1959 indicates the relatively undeveloped status of correlation studies in this part of the spectrum. Many of the gaps are now being filled by the work of Bentley and others, only parts of which have been published at this writing.

Data collections for each of these regions are rather small at present.

F. Accuracy and Precision of Spectra

It is rather easy to check the wavelength calibration of a spectrophotometer with standard samples (e.g., didymium glass in the ultraviolet–visible, polystyrene films or ammonia gas in the infrared) or with line emission light sources. Both wavelength and absorbance standardization for the 200–1000 mμ range have been described in a Bureau of Standards circular,[21] and the use of a copper screen as a universal absorbance standard as well as general calibrating methods are described in spectroscopy texts.[22]

The evaluation of the operators of instruments is a more difficult problem. Early spectrophotometric investigations were difficult technical problems and often attracted workers of unusual experimental talents, but in recent years a spectrum has degenerated to the category of a melting point in the minds of some chemists and may be determined with as little care. Certainly there are a large number of careless errors in recently published spectra. A recent survey of all the ultraviolet–visible data published in 1958–1959 found that replicate determinations

[17] R. F. Goddu in *Advan. Anal. Chem. Instr.* **1**, 347 (1960).

[18] O. H. Wheeler, *Chem. Revs.* **59**, 629–666 (1959).

[19] W. Kaye, *Spectrochim. Acta* **6**, 257–287 (1954); **7**, 181 (1955).

[20] E. D. Palik, *J. Opt. Soc. Am.* **50**, 1329 (1960).

[21] K. S. Gibson, "Spectrophotometry (200 to 1000 millimicrons)." N. B. S. Circular 484, U. S. Government Printing Office, 1949.

[22] R. P. Bauman, "Absorption Spectroscopy," p. 150. Wiley, New York, 1962.

in different laboratories seldom disagreed by more than 1–2 mµ, but fairly often disagreed in molar absorptivity[23] by as much as 20%.

A simple illustration of the sins of omission and carelessness is the ultraviolet spectrum of phenol (see Table I-6), reported in the literature at least 43 times during the 1946–1955 period and certainly many times before and after. Omission of significant parts of the spectrum are common; the statement that a green solution has an absorption maximum at 280 mµ need not be false, but certainly there is more to be said.

TABLE I-6

THIRTEEN VERSIONS OF THE ULTRAVIOLET SPECTRUM OF PHENOL[a]

Solvent	Maxima, mµ (log ϵ)	Date of work
Water	210(3.78), 268 (3.18), 273s(3.10)	1959 (Dearden and Forbes)
	210(3.77), 270 (3.19)	1951
	270 (3.16)	1955
	269.5(3.26)	1950
	269 (—)	1927 (Dahm)
Ethanol	219–25(3.00–3.68), 272–5(2.86–3.43)	1951
	272 (3.27)	1952
	273 (3.38)	1951
	275 (3.32)	1949
	273 (3.27)	1947
Hexane	265(3.15), 275 (3.30), 280(3.25)	1954
	275 (3.30)	1946
	270f(2.30)	1945 (Braude)

[a] All data without author's name are from Vols. I and II of "Organic Electronic Spectral Data" (M. J. Kamlet and H. E. Ungnade, eds.); s is for shoulder; f for fine structure.

Occasionally an article offers a sufficient number of spectra of closely related compounds to allow the reader to evaluate the reliability of the work. In the ultraviolet, for example, there is seldom a good reason for large variations in successive members of an homologous series. Thus (to cite a published example), if the molar absorptivities of a number of 9-alkylguanidines at their constant wavelength of maximum absorption average 8370 with a standard deviation of 540, the reported ϵ of the 9-propyl derivative as 10,200 is implausible, and it is justifiable to surmise that this compound was impure, or that the solution suffered deterioration, or that an experimental mistake was made.

Evaluations of molar absorptivity data in the infrared are rarely made

[23] J. P. Phillips, *Anal. Chem.* **34,** 171 (1962).

because instruments of high resolution are necessary to give true peak heights. The areas under the peaks are less affected by unsatisfactory resolution and have been recommended as more satisfactory than molar absorptivity.[24] The following table (Table I-7) gives standard deviations

TABLE I-7

PRECISIONS OF FREQUENCY AND MOLAR ABSORPTIVITY IN INFRARED SPECTRA[a]

Class of compound	Type of vibration[b]	ν (cm^{-1})	s (cm^{-1})	ϵ	s_ϵ
Esters	C=O	c.1730	—	495	172
Acids, aliphatic	C=O	1706	1.7	411	72
Acids, aromatic	C=O	1695	13.5	523	150
Acid salts	C—O asymmetric	1580	14.9	745	245
Ketones	C=O	c.1674	—	437	127
Amides, primary aliphatic	C=O	1669	8.6	262	45
	NH$_2$ deformation	1631	10.9		
Amides, primary aromatic	C=O	1676	10.9	476	119
	NH$_2$ deformation	1622	9.7		
Amides, secondary aliphatic	C=O	1648	4.5	402	28
Acetanilides	C=O	1673	9.2 ⎫	580	183
Benzanilides	C=O	1658	6.0 ⎭		
Amides, secondary aliphatic	N—H bending	1558	7.6 ⎫		
Amides, secondary aromatic	N—H bending	1539	11.3 ⎭	298	114
Amides, tertiary	C=O	1658	17.3	453	62
Nitro compounds	NO$_2$ symmetric	1350	6.8	484	160
	asymmetric	1532	13.4	477	119
Sulfones	SO$_2$ symmetric	1153	5.4	570	277
	asymmetric	1309	13.4	500	156
Sulfonamides	SO$_2$ symmetric	1161	11.9	491	206
	asymmetric	1326	14.2	316	90
Sulfonic acids	S—O	1196	15.6		
	S—O	1049	12.4	330	89

[a] All data are from reference 24. About 20 compounds of each type were used to obtain s, standard deviation.

[b] Stretching frequencies except as noted.

for both frequency and molar absorptivity for a variety of representative functions from recent work; it is clear that frequencies can be accurate to a few wavenumbers but the absorptivities are often highly variable.

[24] M. St. C. Flett, *Spectrochim. Acta* **18**, 1537 (1962).

G. Solvent, Phase, and Temperature Effects

The general effects of these variables may be crudely stated as follows: polar solvents affect spectra more than nonpolar solvents do; gas phase spectra are more detailed because free from molecular interactions than solution or solid spectra; high temperatures blur and broaden absorption bands while low temperatures sharpen them and permit study of fine structure. The discussion below elaborates on each of these topics.

1. SOLVENTS

In the far ultraviolet the choice of solvent is generally dictated by transparency limitations and only saturated hydrocarbons like hexane are satisfactory.

In the near and far infrared regions the prime factor aside from the presence of absorption bands in the solvent is usually the capacity of the solvent to hold the high concentrations necessitated by the weakness of absorption of most compounds. All solvents with C—H or O—H linkages lack full transparency in the near infrared. Carbon tetrachloride is overwhelmingly preferred in this region.

Infrared spectra are considered little influenced by solvent except for the very large number of compounds in which hydrogen bonding effects must be considered. For these compounds (phenols, alcohols, acids, etc.) detailed discussion is provided in later chapters. Even carbonyl compounds interact with a solvent such as chloroform to yield changes in spectrum.[25] Any association of solute with solvent can be expected to produce some frequency shift. An increase in solvent polarity normally lowers stretching frequencies.

Spurious bands in infrared spectra are often caused by solvent impurities, particularly water; a list of common sources of spurious bands has been assembled.[26] No single solvent is transparent throughout the infrared and since double beam spectrophotometers will not give usable measurements over strong solvent absorptions, it is desirable to know the regions of transparency for available solvents. Charts giving this information are available from chemical supply firms.

Probably the most detailed studies of solvent effects have been made in the ultraviolet–visible region. These effects have been classified into functions of polarity and refractive index of the solvent and actual chemical interactions with it. Nonpolar solvents such as the hydrocarbons

[25] K. B. Whetsel and R. E. Kagarise, *Spectrochim. Acta* **18**, 315, 341 (1962).
[26] P. J. Launer, *Perkin-Elmer Instrument News* **13**, No. 3, 10 (1962).

allow considerable fine structure to be preserved in the spectra of many compounds, while the polar solvents such as water and the alcohols allow only broad and relatively featureless bands. Kundt's rule (which has a great many deviations from it) states that absorption maxima shift to longer wavelengths with increasing refractive index of the solvent. Current theory assigns a larger role to dielectric constant or polarity measures derived from it than to refractive index.

Rather complex equations giving solvent effects in terms of both refractive index and dielectric constant have been derived.[27,28] As an example, McRae's equation for the rather large solvent shifts of the phenol blue spectrum may be cited:

$$\Delta\nu(\text{cm}^{-1}) = (AL_0 + B')(n^2 - 1)/(2n^2 + 1) + C\left(\frac{D-1}{D+2} - \frac{n^2-1}{n^2+2}\right)$$

in which the frequency shift, $\Delta\nu$, is given in terms of dielectric constant D and refractive index n and the weighted wavelength L_0 and three constants A, B', and C. In spite of the generality of this equation it is obviously more than a little difficult to use.

Others[29,30] have attempted to generalize the solvent effect in semi-empirical ways, selecting a peculiarly sensitive absorption band of a single compound as a reference and sometimes devising new polarity measures more satisfactory than dielectric constant.

Shifts of absorption bands as a function of solvent polarity may serve a diagnostic purpose on occasion since some types of electronic transitions are affected differently from others. For example, the $n \rightarrow \pi^*$ transitions of the nonbonding electrons on a nitrogen or oxygen are nearly always accompanied by a shift of the absorption band to shorter wavelengths with increasing solvent polarity, but many $\pi \rightarrow \pi^*$ transitions have a bathochromic shift under the same conditions. In an α,β-unsaturated aldehyde or ketone the long wavelength band is $n \rightarrow \pi^*$ and the stronger short wavelength band $\pi \rightarrow \pi^*$; in polar solvents such as water the spacing of these two bands is less than in nonpolar solvents because of the opposing solvent action on the two transitions (see Table I-8).

Aside from actual chemical reactions between solute and solvent the

[27] E. G. McRae, *J. Phys. Chem.* **61**, 562 (1957); *Spectrochim. Acta* **12**, 192 (1958).
[28] O. E. Weigang, *J. Chem. Phys.* **33**, 892 (1960).
[29] E. M. Kosower, *J. Am. Chem. Soc.* **80**, 3253 (1958).
[30] W. M. Schubert, H. Steadly, and J. M. Craven, *J. Am. Chem. Soc.* **82**, 1353 (1960).

largest effects of solvent are encountered in those compounds that engage in charge-transfer with the solvent; i.e., excited electrons in the compound may in effect occupy in part unfilled orbitals of the solvent, or vice versa. (Sometimes this effect has been considered to produce an actual stable complex.) The first systematic study of bands of this

TABLE I-8

EFFECT OF SOLVENT POLARITY ON WAVELENGTHS OF CARBONYL GROUP BANDS[a]

		Wavelength (mμ)				
		Mesityl oxide		Crotonaldehyde		Acetone
Solvent	Dielectric constant	$\pi \to \pi^*$	$n \to \pi^*$	$\pi \to \pi^*$	$n \to \pi^*$	$n \to \pi^*$
Hexane	2	229.5	327	217	321	279
Ether	4	230	326			
Chloroform	4.8	238	315			277
Butanol	18	237	311			
Ethanol	25	237	310	220	318	272
Methanol	33	237	309			270
Water	81	243	305	220	310	264.5

[a] All data are from Vols. I and II of "Organic Electronic Spectral Data" [M. J. Kamlet (Vol. I) and H. E. Ungnade (Vol. II) eds.]. Interscience, New York, 1960.

kind was performed on solutions of iodine in various hydrocarbons, but the most spectacular effects so far observed come from combinations of tetracyanoethylene with various hydrocarbons. The extent of charge-transfer varies greatly from one hydrocarbon to another, and the colors of tetracyanoethylene solutions correspondingly range over the entire visible spectrum (see Table I-9). The capacity of dissolved oxygen to show the charge-transfer effect with alcohols, hydrocarbons and other organic compounds accounts for a number of spurious bands in the published ultraviolet spectra of such compounds.[31] Many of the characteristic colors of inorganic ions are accounted for by charge-transfer (see Chapter VII).

Of the definite chemical reactions between solute and solvent ionization produces changes in spectra that may be highly characteristic as well as easy to analyze. For any substance with even the slightest acid or base properties spectra in acid, base and neutral solutions are minimum requirements. Structure determination in such compounds as the purines, pyrimidines, and nucleic acids that have several acid or base

[31] H. Tsubomura and R. S. Mulliken, J. Am. Chem. Soc. 82, 5966 (1960).

functions and thus several changes of spectrum over the available pH range may be considerably facilitated by spectra at 1–2 pH unit intervals. It is obviously possible to use spectra to determine ionization

TABLE I-9

CHARGE-TRANSFER BANDS OF TETRACYANOETHYLENE WITH HYDROCARBONS[a]

Hydrocarbon	Maximum (mμ)	Hydrocarbon	Maximum (mμ)
Benzene	385	Biphenyl	500
Naphthalene	560	Phenanthrene	540
Azulene	740	Picene	590
1,2-Benzanthracene	748	Chrysene	630
3,4-Benzotetraphene	800	Pyrene	720
3,4-Benzopyrene	820	Perylene	920

[a] Data are from spectra in chloroform by M.J.S. Dewar and H. Rogers, *J. Am. Chem. Soc.* **84**, 395 (1962).

constants and the spectrophotometric methods may indeed be applied to a wider range of acid strengths than any other known method. In Chapter III some common techniques for determining ionization constants are described in the section on acids.

2. PHASE

Vapor phase spectra are more richly endowed with fine structure than solution or solid spectra, an effect clearly related to the relative freedom of gas molecules from interactions with other molecules and the resulting partial transfer of energy.

In ultraviolet–visible spectra there is a general shift to longer wavelengths in passing from vapor phase to solution spectra. If an ultraviolet spectrum for the gas phase can be obtained, the very large number of bands for most compounds makes characterization extremely easy,[32] and bibliographies of spectra of organic vapors have been compiled.[33] Among the compounds thus determined are benzene, halogenated benzenes, toluene, anilines, phenols, pyridines, nitriles, and some aldehydes and ketones.

Probably the most common problem imposed by change of state is encountered in the comparison of infrared spectra in potassium bromide

[32] R. W. B. Pearse and A. G. Gaydon, "The Identification of Molecular Spectra." Chapman and Hall, London, 1950.

[33] W. D. McGrath, W. F. Pickering, R. J. Magee, and C. L. Wilson, *Talanta* **9**, 227 (1962).

pellets with solution spectra. The solid state spectra may differ from the solution spectra in lacking absorption bands from the less stable of a pair of rotational isomers that can exist together in a solution but not in an oriented solid. New bands may be added to the solid spectrum because, for example, a vibrational mode may be split into in-plane and out-of-plane components in the solid but not in a solution. There are additional possible effects from chemical reactions with potassium bromide, solution of the sample in the solid bromide, polymorphic forms, specific orientations with pressure, and other factors.[34,35] It is particularly difficult to obtain pellets free from moisture.

As a general rule the increase of molecular interactions from vapor to liquid to solid phases is associated with a decrease in stretching frequencies in the infrared and an increase in deformation frequencies. Except where hydrogen bonding is involved, the magnitude of these shifts is not large.

With suitable instrumentation ultraviolet–visible and infrared spectra with polarized light can be obtained, and these methods help to establish orientation and molecular structure in solid samples that are fibers, partially crystalline or otherwise capable of differing absorption in differing directions along the molecular axes.

3. TEMPERATURE

A high temperature affects electronic transitions by broadening the range of energies able to produce a given transition, with the usual result that the absorption band has lower peak intensity but broader wavelength span than at lower temperature. Exceedingly low temperatures often yield much sharper and more detailed spectra, and theoretical investigations are often facilitated by spectra at liquid air temperatures on solutions that are solid glasses at such low temperatures but noncrystalline.

In some instances heat may produce a radical change of color, a phenomenon called thermochromism.[36] Sometimes this effect is simply the broadening by heat of an ultraviolet band until it edges into the visible, but the more prominent examples involve a reversible reaction in which the position of equilibrium is sharply altered by temperature. An even more exotic effect is photochromism, in which illumination of the sample changes its color and heat or illumination with a different

---

[34] V. C. Farmer and J. D. Russell, *Spectrochim. Acta* **18**, 461 (1962).

[35] G. Duyckaerts, *Analyst* **84**, 201 (1959).

[36] J. E. Day, *Chem. Revs.* **63**, 65 (1963).

wavelength band erases the change. Compounds capable of this generally require rather low temperatures for the effect,[37] and free radical formation is often the best explanation of the observed changes.[38]

The influence of temperature on infrared spectra has been less thoroughly studied, probably because of sample handling difficulties over a wide range of temperature, but it does not appear to exert any very striking effect unless there is a change of phase.[39]

[37] E. Berman, R. E. Fox, and F. D. Thomson, *J. Am. Chem. Soc.* **81,** 5605 (1959).
[38] R. B. Woodward and E. Wasserman, *J. Am. Chem. Soc.* **81,** 5007 (1959).
[39] M. P. Lisitsa and I. N. Khalimonova, *Optics and Spectry.* **11,** 179 (1961).

II

Hydrocarbons

Since some portions of the spectra of the parent hydrocarbons are preserved in nearly all other kinds of organic compounds, they must be considered first in any systematic discussion of spectra. More spectra of hydrocarbons in all regions are available than of any other class of compound, and much success has attended the numerous theoretical analyses of these data.

In this and following chapters the following conventions of nomenclature are used. Wavelength of maximum absorption is indicated by λ_{max}, frequency by ν, usually in wavenumbers. Fine structure in ultraviolet maxima is symbolized by f, shoulders or inflections by s. In infrared spectra (s), (m), (w), and (v) stand respectively for strong, medium, weak, and variable intensity bands. A bathochromic shift is a change of maximum to a longer wavelength and a hypsochromic shift a change to shorter wavelength; a hyperchromic effect is an increase in molar absorptivity, ϵ, and a hypochromic shift is a decrease.

A. Acyclic Saturated Hydrocarbons

1. Far Ultraviolet

There is general transparency above 170 mμ for the pure hydrocarbons,[1] and the simpler liquid members are therefore good solvents for the far ultraviolet and ultraviolet spectra of other compounds. In the vacuum ultraviolet the longest wavelength bands of methane and ethane vapors are at 122 and 135 mμ respectively,[2] but highly branched hydrocarbons[3] have considerably higher λ_{max} (e.g., 2,2,4-trimethylpentane, 154 mμ, $\epsilon \sim 10,000$). Turner[4] has summarized the ionization potentials

[1] W. J. Potts, Jr., *J. Chem. Phys.* **20**, 809 (1952).

[2] G. Moe and A. B. F. Duncan, *J. Am. Chem. Soc.* **74**, 3140 (1952).

[3] D. W. Turner, *Chem. Ind. (London)* **1958**, 626.

[4] D. W. Turner, *in* "Determination of Organic Structures by Physical Methods" (F. C. Nachod and W. D. Phillips, eds.), Vol. II, pp. 339–400. Academic Press, New York, 1962.

in electron volts (1 eV is equivalent to 8066 cm^{-1}) for a great many hydrocarbons from vacuum ultraviolet data.

2. ULTRAVIOLET–VISIBLE

Since the only possible electron transitions are $\sigma \rightarrow \sigma^*$ and these are in the far ultraviolet, there is complete transparency in the ultraviolet–visible.

3. NEAR INFRARED

The strongest bands of saturated hydrocarbons below 2.5 μ are generally combination bands (see Chapter I) in the 2.3–2.45 μ region. The C—H stretching fundamental is near 3.4 μ in the infrared, and the first four overtones are calculated by elementary arithmetic to fall near 1.7, 1.1, 0.9, and 0.75 μ. Hydrocarbons do show bands near these locations, with the 1.7 μ first overtone the strongest and most generally useful.[5] There is progressive decay of intensity to the fourth overtone near 0.75 μ which is so weak that a 10 cm thickness of undiluted sample may be needed to detect it at all.

There are useful differences in the first two overtones for methyl and methylene groups; thus the first overtone for CH_3 appears as a triplet at 1.78, 1.70, and 1.69 μ, but for the CH_2 group at 1.76, 1.73, and 1.70 μ.[6] The second overtone has been claimed to be best for estimating methyl to methylene rations because the methyl maximum lies at 1.195 μ and the methylene at 1.215 μ and few interferences from other groups are in this region. The ratios are computed on the assumption that the peak area or height is proportional to the number of methyls or methylenes present per molecule.

Many hydrocarbons also have combination bands near 1.4 μ. A small collection[7] of representative near infrared spectra includes several simple hydrocarbons.

4. INFRARED

There are C—C bands in all saturated hydrocarbons except methane but they are usually weak and in any case of little characterization value because of their universal occurrence. Consequently the most valuable frequencies are C—H stretching vibrations near 2900 cm^{-1} and deformation frequencies near 1400 cm^{-1}, together with a few skeletal vibrations

[5] R. F. Goddu, *Advan. Anal. Chem. Instr.* **1**, 347 (1960).

[6] W. Kaye, *Spectrochim. Acta* **6**, 257 (1954).

[7] Beckman Instruments, Collection of 132 Near Infrared Spectra. Fullerton, Cal., 1959.

at smaller frequencies for such specific structural features as isopropyl, *tert*-butyl, and polymethylene groups.

The asymmetric and symmetric stretchings of the methyl group lie at 2962 and 2872 cm⁻¹ respectively and very rarely deviate more than 10 cm⁻¹ from these values even in complex compounds.[8] The corresponding methylene bands may be located with equal precision at 2926 and 2853 cm⁻¹. All these bands are strong, with intensities proportional to the number of methyl or methylene groups in the compound. By contrast a tertiary C—H stretching is represented by a weak band at 2890 cm⁻¹. These and other C—H stretching values have been summarized in some detail (see Table II-1).

TABLE II-1

LOCATION OF C—H STRETCHING FREQUENCIES IN THE INFRARED[a]

Group or class	Absorption maxima (cm⁻¹)
Methyl	2952–2972; 2862–2882
in sulfur or oxygen compounds	2955–2992; 2867–2897
Methylene	2916–2936; 2843–2863
in sulfur or oxygen compounds	2922–2948; 2846–2878
Cyclopropanes	3072–3100; 2995–3033
Epoxides	3029–3058 or 2990–3004
Cyclobutane CH₂ groups	2977–2999; 2874–2924
Cyclobutane CHR groups	2855–2874
Cyclopentanes	1952–2959; 2853–2866
Olefinic =CH₂	3077–3092
Olefinic =CHR	3012–3025
Aldehyde C—H (aliphatic)	2810–2830; 2710–2724
Aldehyde C—H (aromatic)	2818–2850; 2720–2745
Methoxy	2815–2832
Aromatic C—H	3048–3096; 3025–3039; 3000–3020

[a] All data are transcribed from a graph in reference 8.

The C—H bending frequencies for these groups are as follows: CH₃ (attached to carbon) at 1450 ± 20 (m)(asymmetrical) and 1380–1370 cm⁻¹(s)(symmetrical); methylene at 1465 ± 20 cm⁻¹(m); and C—H at 1340 cm⁻¹(w). A methyl group attached to an element other than carbon may undergo a considerable alteration in these frequency assignments, however.

Considerable interest by polymer chemists in polymethylene chains has led to very detailed analysis of vibrational bands associated with

[8] S. E. Wiberley, S. L. Bunce, and W. H. Bauer, *Anal. Chem.* **32**, 217 (1960).

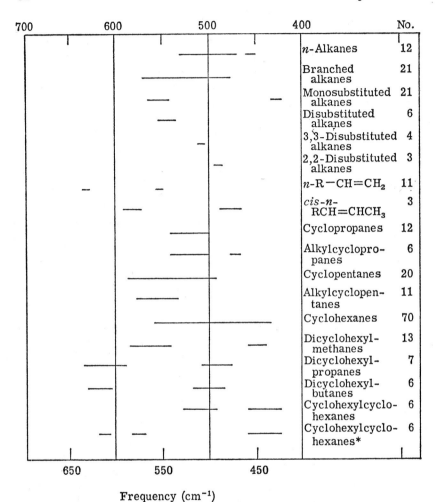

Frequency (cm⁻¹)

FIG. II-1. Far infrared spectra of alkanes and alkenes (238 compounds). An asterisk indicates the high-boiling members of the series. Adapted from reference 11.

the —CH_2— group.[9] The simplest compound with this group is propane, and it has six vibrational modes for the methylene, at 2926, 2853, 1460, 1336, 1278 and 748 cm⁻¹. The first two are the asymmetric and symmetric stretchings mentioned above, the third a bending frequency, and the last three (in order) wagging, twisting, and rocking frequencies. Longer chains will have additional C—C stretching modes, C—C—C angle

[9] N. Sheppard in *Advan. Spectry.* **1,** 288 (1959).

TABLE II-2
PRINCIPAL MAXIMA IN THE FAR INFRARED (15–35 μ) SPECTRA OF HYDROCARBONS[a]

Compound	Maxima (μ)
Pentane	21.3, 26(w)
Heptane	20.6(s), 22.0(w), 24.5(w)
Decane	19.5, 21.9, 24.5(broad)
Hexadecane	18.7, 22.0(w)
2-Methylpentane	18.3, 20.6, 22.5, 26.1
2-Methylheptane	17.9, 18.3, 20.3(s), 22.2, 23.5, 24.6, 26.1
2,2-Dimethylbutane	20.7(s), 24.0, 24.5, 27.7(w)
2,3-Dimethylpentane	18.2(s), 20.9, 21.7, 23.3
3,3-Diethylpentane	18.0, 19.9(s), 22.0, 24.6
2-Cyclopropylpropane	15.1, 19.1, 21.5(s), 27(w), 29.5, 32.5(w)
Vinylcyclopropane	15.4(s), 19.0, 22.1(s), 31.2
2-Cyclopropylbutane	15.2, 18.3, 19.0, 19.8, 21.4(s), 22.5(s), 24.6, 27.0
Cyclopentane	18.5(s)
Methylcyclopentane	15.5(broad), 18.9(s), 23.3
Isopropylcyclopentane	16.3, 18.3(s), 21.5, 22.8, 24.0, 27.6
Cyclohexane	19.0(s)
Methylcyclohexane	15.0, 16.5(s), 18.4(s), 21.0, 22.5(s), 24.6
1-Pentene	16.0(s), 18.1(s), 22.9, 25.5
1-Nonene	15.9(s), 18.1(s), 19.6(w), 22.5
2,3-Dimethyl-1-butene	18.4(s), 20.5(s), 22.9, 23.2(w), 32.2
cis-2-Pentene	17.4(s), 21.5(s)
trans-2-Pentene	17.4(w), 20.6(w), 24.4(s), 33.5–34.0
cis-2-Hexene	17.3(s), 20.9
trans-2-Hexene	18.9(w), 20.5(w), 22.5, 25.8, 31.5(broad)

[a] Data are estimated from graphs in reference 11.

bending modes and torsional vibrations about internal C—C bonds. According to Bellamy[10] straight chains containing four or more linked methylenes are characterized by a strong band at 720–750 cm⁻¹. (See Chapter VIII for a few remarks concerning the infrared spectrum of polyethylene.)

Isopropyl groups are characterized by a splitting of the symmetrical CH_3 bending into two strong equal bands at 1380–1385 and 1365–1370 cm⁻¹, as well as by skeletal vibrations at 1170 ± 5 (s) and 1140–1170 cm⁻¹.

Tertiary butyl groups show the methyl bending as a medium band at

[10] L. J. Bellamy, "The Infrared Spectra of Complex Molecules," 2nd ed. Wiley, New York, 1958.

1385–1395 and a strong band at 1365 cm^{-1}; there is also a strong 1250 $\pm$ 5 cm^{-1} band and another in the 1200–1250 cm^{-1} range.

5. FAR INFRARED

In the cesium bromide region from 15 to 35 μ bands in both the 18.6–20.6 and 21.8–22.1 μ regions have been observed for all the straight chain hydrocarbons from hexane to hexadecane.[11] Branched alkanes tend to exhibit more and stronger bands, largely in the 17.5–20.6 μ range. For fifty alkanes of all types a broad increasing absorption from 30 to 35 μ probably originates in the planar skeletal vibrations of the zigzag carbons in the chain. Although a chart of structural generalizations has been published (see Fig. II-1), examination of individual compound spectra (Table II-2) suggests that any generalizations for this part of the spectrum are risky.

B. Cyclopropanes

1. FAR ULTRAVIOLET

The electronic spectra of the cyclopropanes are closer to those of olefins than to higher saturated rings. Cyclopropane gas has a dozen weak bands in the 182–189 mμ region [12] and these are seen in the solution spectrum as a weak band at 190–195 mμ.[13] The gas spectrum also has bands in the vacuum ultraviolet near 120, 145 and 159 mμ.

2. ULTRAVIOLET–VISIBLE

Simple cyclopropanes should be transparent, though suitable auxochromic substituents may shift the 190 mμ band into the ultraviolet. There is evidence that cyclopropyl groups may be conjugated with double bonds to yield compounds with spectra intermediate in wavelength of maximum absorption between alkenes and dienes, and a similar conjugation with a carbonyl double bond has also been demonstrated.[14]

3. NEAR INFRARED

The first overtone of the C—H stretching frequency is near 1.64 μ and the second at 1.1 μ, but the strongest band in cyclopropanes is often a combination near 2.24 μ.[15] In substituted cyclopropanes the 1.64 μ

---

[11] F. F. Bentley and E. F. Wolfarth, *Spectrochim. Acta* **15**, 165 (1959).

[12] P. Wagner and A. B. F. Duncan, *J. Chem. Phys.* **21**, 516 (1953).

[13] A. Pullman and B. Pullman, *Discussions Faraday Soc.* **9**, 46 (1950).

[14] E. M. Kosower, *J. Am. Chem. Soc.* **80**, 3261 (1958).

[15] W. H. Washburn and M. J. Mahoney, *J. Am. Chem. Soc.* **80**, 504 (1958).

band intensity may be used to determine the number of CH_2 groups left in the ring, because the intensity is twice as great for two unsubstituted methylenes as for one.[16] Identification of the cyclopropyl ring through this band is said to be superior to the use of any infrared absorption; some typical values are: ethylcyclopropane, 1.639; *n*-butylcyclopropane, 1.639; 1,1,2,2-tetramethylcyclopropane, 1.650; bicyclopropyl, 1.638; dicyclopropylmethane, 1.638; cyclopropyl carbinol, 1.637; and 1,1-dichlorocyclopropane, 1.625 μ.

4. INFRARED

In a group of more than sixty substituted cyclopropanes Wiberley[8] found C—H stretching bands at 3099–3072 and also 3033–2995 cm^{-1}. A medium intensity cyclopropane ring vibration at 1020–1010 cm^{-1} and a less useful[17] ring deformation near 868 cm^{-1} are other features,[9,18] and a reasonably reliable identification of the cyclopropyl ring may require examination of all three of these regions.

5. FAR INFRARED

Although cyclopropane itself lacks characteristic absorption in the 15–35 μ region, several cyclopropyl derivatives are characterized by a 18.5–20.0 μ band, and some alkylcyclopropanes have a strong maximum at 21.2–21.6 μ (see Table II-2). Only a few spectra in this region have been reported.

C. Cyclobutanes, Cyclopentanes, and Cyclohexanes

1. FAR ULTRAVIOLET AND ULTRAVIOLET–VISIBLE

Since the strain that gave the cyclopropanes some unsaturated character is much less in the higher ring systems, the latter are ordinary saturated hydrocarbons transparent down to 175 mμ.[19]

2. NEAR INFRARED

Small differences in the C—H overtones of methylene groups in cyclopentanes and cyclohexanes have been noted;[20] cyclopentanes have the

---

[16] P. G. Gassman, *Chem. Eng. News* April 9, 1962, p. 49.

[17] S. A. Liebman and B. J. Gudzinowicz, *Anal. Chem.* **33**, 941 (1961).

[18] C. F. Wilcox, Jr., and R. R. Craig, *J. Am. Chem. Soc.* **83**, 3866 (1961).

[19] L. W. Pickett, M. Muntz, and E. M. McPherson, *J. Am. Chem. Soc.* **73**, 4862 (1951).

[20] A. Evans, R. R. Hibbard, and A. S. Powell, *Anal. Chem.* **23**, 1604 (1951).

normal 1.215 μ peak plus an additional sharp band at 1.195 μ, and the cyclohexane bands are somewhat broader than is usual for this overtone.

3. INFRARED

The methylene stretching frequencies in cyclobutanes are higher than in cyclopentanes and both are above the normal methylene values (see Table II-1). In a number of fused ring and cage structures where methylene hydrogens are in very close proximity abnormally high stretching frequencies have also been reported.[21]

The CH_2 group in cyclobutanes has stretching frequencies of 2999–2977 and 2924–2878 cm⁻¹, while a CHR (R is alkyl) group absorbs at 2974–2855 cm⁻¹ (Table II-1). In cyclopentanes the methylene frequencies are 2959–2952 and 2866–2853 cm⁻¹.

According to Willis and Miller[22] skeletal frequencies for the cyclic hydrocarbons are located as follows: cyclobutane, 1020 and 950; cyclopentane, 990 and 890; and cyclohexane, 1020 and 800 cm⁻¹. Bellamy[10] lists ring frequencies for cyclobutane at 910 cm⁻¹, for cyclopentanes near 977 (but with wide individual variations), and for cyclohexanes in the 1005–925 and 1055–1000 cm⁻¹ regions.

4. FAR INFRARED

Ring deformations at 630 for cyclobutane, 542 for cyclopentane and 532 cm⁻¹ for cyclohexane have been reported. Simple alkyl derivatives of cyclopentane and cyclohexane have much more complex spectra than the parent hydrocarbons in this region (Table II-2).

D. Simple Olefins

1. FAR ULTRAVIOLET

The single double bond is characterized by a strong absorption below 200 mμ, and indeed the most important single use of far ultraviolet spectra has been associated with the analysis of this band, particularly in steroids. It is a broad, intense absorption ($\epsilon \sim 10,000$) near 175 mμ for the simpler olefins (see Table II-3) and results from a $N \rightarrow V$ transition of a π electron. In vapor phase spectra there is pronounced fine structure and interpretation is made difficult by the existence of as many

[21] D. Kivelson, S. Winstein, P. Bruck, and R. L. Hansen, *J. Am. Chem. Soc.* **83**, 2938 (1961).

[22] G. H. Beaven, E. A. Johnson, H. A. Willis, and R. G. J. Miller, Molecular Spectroscopy, p. 229. Macmillan, New York, 1961.

as three overlapping absorption processes in the 160–200 mμ interval. Provided that the double bonds are insulated from each other by two or more methylenes, diolefins have the same λ_{max} but a doubled molar absorptivity.

TABLE II-3

FAR ULTRAVIOLET MAXIMA IN THE SPECTRA OF SIMPLE OLEFINS[a]

Compound	Solvent or state	Principal maxima mμ (log ϵ)
Ethylene[b]	Vapor	162[c]
Propene	Vapor	173f (4.2)
1-Butene	Vapor	175f (4.2)
1-Pentene	Vapor	177f (4.2)
1-Hexene	Vapor	179f (4.0)
1-Octene	Heptane	177 (4.1)
Cyclopropene[d]	Vapor	under 185
Cyclobutene[d]	Vapor	175 (4.1), 191 (3), 195 (2.9)
Cyclopentene[d]	Vapor	183 (4.1), 200s
Cyclohexene	Vapor	182 (3.9)
cis-2-Butene	Vapor	175 (4.3)
trans-2-Butene	Vapor	177 (4.1)
cis-2-Pentene	Vapor	177 (4.3)
trans-2-Pentene	Vapor	181 (4.2)
cis-2-Hexene	Vapor	178 (4.2)
	Heptane	183 (4.1)
trans-2-Hexene	Vapor	179 (4.1)
	Heptane	184 (4.0)
cis-2-Octene	Heptane	183 (4.1)
trans-2-Octene	Heptane	179 (4.1)
2-Methyl-1-propene	Vapor	188 (4.1)
2-Methyl-1-butene	Vapor	188 (4.0)
2-Ethyl-1-butene	Vapor	187 (3.9)
2-Methyl-1-pentene	Vapor	189 (3.9)

[a] Data are from "Organic Electronic Spectral Data" [M. J. Kamlet (Vol. I) and H. E. Ungnade (Vol. II), eds.]. Interscience, New York, 1960, except as shown.

[b] A complex spectrum; see reference 4.

[c] P. G. Wilkinson and R. S. Mulliken, *J. Chem. Phys.* **23,** 1895 (1955).

[d] From reference 27.

Turner[4] has reached the interesting conclusion that stray light errors are so prevalent in far ultraviolet work and the spectra so complex that correlations of double bond substitution patterns should be based not on λ_{max} but on the wavelength required to reach a molar absorptivity of 1000 (approaching from the long wavelength side). Obviously these values are longer wavelengths than the maxima and will consequently fall in a region of lesser experimental difficulty. The more substituents on the double bond the higher this ϵ_{1000} will be; in a group of 47 examples the majority have ϵ_{1000} above 210 mμ and thus actually in the ultraviolet. One substituent on the double bond makes ϵ_{1000} about 196 mμ; two, 202–211; three, 213–221.5; and four, 223–229 mμ.

Earlier work[23] was, perhaps mistakenly, more positive about the value of λ_{max} as a guide to substituent positions and geometric isomerism. Thus, 1-alkenes should have λ_{max} 175 $\pm$ 2.5 ($\epsilon \sim$ 11,800); 2-alkyl-1-alkenes, λ_{max} 187.5 $\pm$ 1.5($\epsilon \sim$ 8900); *cis*-2-alkenes, λ_{max} 176.5 $\pm$ 1.5 ($\epsilon \sim$ 12,300); and *trans*-2-alkenes, λ_{max} 179 $\pm$ 1 mμ ($\epsilon \sim$ 11,700). As Table II-3 shows, these claims may be more precise than is currently justified.

In steroids the number of substituents on an isolated double bond has been stated[24] to have the following relationship to λ_{max}: two substituents, 182–184 mμ; three, 188–193; and four, 196–200 mμ. For a group of 46 steroids and triterpenes with a single double bond a maximum in the 193–205 mμ region has been observed.[25]

2. ULTRAVIOLET–VISIBLE

Solution spectra of olefins show little except "end absorption," a rise in absorbancy as the short wavelength limit of the ultraviolet is approached. Substantial end absorption (at about 210 mμ) indicates a potential maximum in the far ultraviolet, and it has been used, though mainly before the advent of far ultraviolet instruments, to determine the degree of substitution on the double bond in the same way as Turner's ϵ_{1000} in the far ultraviolet. The more heavily substituted the double bond the greater the end absorption.[26]

3. NEAR INFRARED

The terminal methylene or $=CH_2$ group has a very distinctive overtone near 1.62 μ in carbon tetrachloride solutions. Goddu[5] lists 27 vinyl compounds with this band between 1.61 and 1.636 μ with an average molar absorptivity about 0.3. Weaker overtones for olefins near 1.1, 0.9, and 0.75 μ are about the same as for any C—H containing compound (see Fig. II-2), but may have more numerous fine structure peaks than is usual for saturated compounds. A combination band has been found

[23] L. C. Jones and L. W. Taylor, *Anal. Chem.* **27,** 228 (1955).

[24] D. W. Turner, *J. Chem. Soc.* **1959,** 30.

[25] P. S. Ellington and G. D. Meakins, *J. Chem. Soc.* **1960,** 697.

[26] P. Bladon, H. B. Henbest, and G. W. Wood, *J. Chem. Soc.* **1952,** 2737.

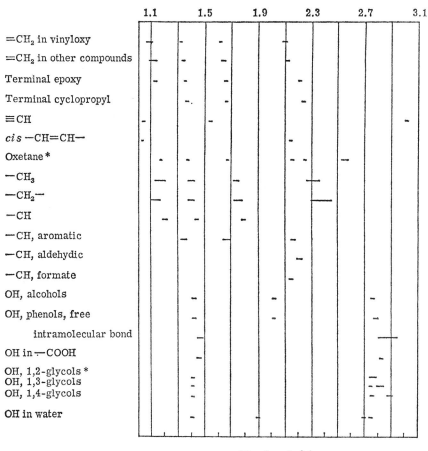

Wavelength (μ)

Fig. II-2. Near infrared absorption bands of C—H and O—H functions. Molar absorptivities are generally very low at the left for each function ($\sim 10^{-2}$) and increase toward the right or longer wavelength side of the chart to values as high as 10–100. Most spectra were determined in carbon tetrachloride solutions. Adapted from R. F. Goddu and Decker, *Anal. Chem.* **32,** 140 (1960). Asterisk indicates that in these cases the longest wavelength interval usually contains two maxima.

for 15 vinyl compounds at 2.09–2.12 μ (0.1–0.6). Compounds with *cis* double bonds have a 2.14 μ peak that is lacking in *trans* isomers.

4. INFRARED

The C=C stretching vibration is generally weak in isolated double bonds, normally falling in the 1660–1640 cm^{-1} range but as much as

20 cm^{-1} higher or lower for some compounds. In tetrasubstituted ethylenes it is very hard to find this weak frequency at all, and its diagnostic value for these compounds is thus negligible.

Substitution effects are more apparent in the C—H stretching frequencies. Compounds containing the vinyl group (—CH=CH$_2$) have a 3077–3092 cm^{-1} band for the methylene and a 3012–3025 cm^{-1} band attributed to the —CH=. Other characteristic absorption for the vinyl group includes medium bands at 1856–1800 and 1420–1410 cm^{-1} (the latter a CH$_2$ in-plane deformation), and strong bands at 995–985 (CH out-of-plane deformation) and 915–905 cm^{-1} (CH$_2$ out-of-plane deformation).

In compounds having the —CH=CH— grouping a distinction between *cis* and *trans* isomers is furnished by a 690 cm^{-1} band in the *cis* compounds, and a pair of bands at 1310–1295 (CH in-plane deformation) and 970–960 cm^{-1} (CH out-of-plane deformation) in the *trans* form.

Trisubstituted ethylenes have a strong band in the 840–790 cm^{-1} region attributed to the =C—H deformation. Though the band position is somewhat variable it has been used to identify this group in a variety of steroids and terpenes. There are also C—H stretching frequencies at 3040–3010 cm^{-1}.

Identification of RR'C=CH$_2$ compounds is aided by the presence of the following bands: 3095–3075 (m), 1800–1750 (m), and 895–885 cm^{-1}(s).

The lower cycloalkenes have the =C—H stretching band displaced in an order roughly indicative of ring size[27] (Table II-4). The C=C band for these compounds is not so helpful in general, though its position and very weak intensity in cyclobutene[28] are distinctive.

TABLE II-4

CHARACTERISTIC INFRARED FREQUENCIES OF THE LOWER CYCLOALKENES[27]

Ring size	=C—H band (cm^{-1})	C=C band (cm^{-1})
3	3076	1641
4	3048, 3126	1566[28]
5	3061	1611
6	3024, 3067	1649
7	3020, 3059	1650
8	3016, 3053	1648

[27] K. B. Wiberg and B. J. Nist, *J. Am. Chem. Soc.* **83**, 1226 (1961).

[28] C. F. Wilcox, Jr., S. Winstein, and W. G. McMillan, *J. Am. Chem. Soc.* **82**, 5452 (1960). A Raman shift.

The above survey of the infrared spectra of alkenes is rather superficial. Bellamy[11] lists 111 articles on this subject, many of them devoted to detailed analysis of the $C=C$ stretching vibration.

5. FAR INFRARED

The unbranched 1-alkenes have strong bands in the 15.7–16.0 and 18.05–18.2 μ regions.[11] Branched 1-alkenes with alkyl substituents on or adjacent to the double bond absorb strongly in the 17.9–18.8 μ range. 4-Alkyl-1-alkenes have a strong band at 16.0–16.3 μ. Considerable differences between *cis* and *trans* isomers are also apparent in this region (Table II-2).

E. Conjugated Olefins

1. FAR ULTRAVIOLET

In contrast to compounds with a single double bond the conjugated dienes generally have an absorption minimum in the 175–190 mμ region.[4] The higher polyenes are virtually free from absorption unless one or more of the double bonds is *cis* rather than *trans*.

2. ULTRAVIOLET–VISIBLE

a. Dienes.

The high intensity band of butadiene at 217 mμ ($\epsilon \sim 20,000$) is shifted about 5 mμ to longer wavelengths by each alkyl substituent and also 5 mμ if the double bond that is part of the conjugated system is exocyclic to a 6-membered ring.[29] This celebrated generalization, part of "Woodward's rules," was formulated in a day of greater confidence in exact spectra-structure relationships in the ultraviolet than the present, but it stands up well and has been extensively applied (with modifications) to problems of steroid structure. A more recent examination of the rule[30] has suggested that a second alkyl substituent should be assigned a somewhat larger bathochromic effect than the first. Tables of examples are given by Jaffé and Orchin.[31] Alcohol is the usual solvent for these determinations, but hydrocarbon spectra are relatively insensitive to solvent changes.

The following catalog (Table II-5) of band maxima as a function of

[29] R. B. Woodward, *J. Am. Chem. Soc.* **63**, 1123 (1941); **64**, 72, 76 (1942).

[30] W. F. Forbes and R. Shilton, *J. Org. Chem.* **24**, 436 (1957).

[31] H. H. Jaffé and M. Orchin, "Theory and Applications of Ultraviolet Spectroscopy," p. 198. Wiley, New York, 1962.

diene type [32] is of about the same vintage as Woodward's rules but a little more elaborate. It may be noted that cyclic dienes absorb at longer wavelength than the acyclic. The value of λ_{max} is greater for the cyclohexadienes than for either larger or smaller rings, the minimum value of λ_{max} apparently in the 9-membered ring.

TABLE II-5

PRINCIPAL ULTRAVIOLET ABSORPTION MAXIMUM IN CONJUGATED DIENES[32]

Diene type	Maximum $(m\mu)$
Acyclic	
no cyclic substituents	217–228
one cyclic substituent	236
two cyclic substituents	246–248
Semicyclic (one double bond in ring, other not)	230–242
Cyclohexadienes	256–265
Bicyclic (with double bonds in different rings)	236
Polycyclic	
double bonds in one ring	260–282
double bonds in different rings	235–249

b. Conjugated Polyenes and Related Systems.

It has been experimentally established for a long time that the wavelength of maximum absorption for the conjugated chain R—(CH=CH)$_n$—R′ increases with the number of units in conjugation, n, but that the rate of increase slows at higher values of n. For small values of n the molar absorptivity increases linearly. Comparable behavior is shown whether the ends of the chain are alkyl or aryl, one end is an aldehyde or acid function, the double bond is replaced by a triple bond, and some of the carbons are replaced by nitrogens. Indeed, any system of alternating single and multiple bonds not containing a charged component may be expected to show this more or less regular bathochromic shift with increasing length. The following is a partial list (with references) of the systems most intensively examined:

$$H—(CH=CH)_n—H \quad [33]$$
$$CH_3—(CH=CH)_n—CH_3 \quad [34]$$

[32] H. Booker, L. K. Evans, and A. E. Gillam, *J. Chem. Soc.* **1940**, 1453.

[33] F. Sondheimer, D. A. Ben-Efraim, and R. Wolovsky, *J. Am. Chem. Soc.* **83**, 1675 (1961).

[34] F. Bohlmann and H. Mannhardt, *Chem. Ber.* **89**, 1307 (1956).

$$Ph—(CH=CH)_n—Ph$$

$$Me_3C—(C\equiv C)_nCMe_3{}^{35}$$

$$CH_3—(CH=CH)_nCHO{}^{37}$$

$$CH_3—(CH=CH)_nCOOH{}^{37}$$

$$R—(CH=CH)_nCH=N—N=CH—(CH=CH)_nR{}^{38}$$

In the $CH_3—(CH=CH)_n—CH_3$ series all members up to $n = 11$ are known (see Table II-6), and allowance of a few structural irregularities for higher values permits the series to include the carotenoids and other yellow to red pigments of natural origin or biochemical significance. Until 1961 the parent series $H—(CH=CH)_n—H$ was known only to $n = 5$ but many additional members have been recently synthesized.[33]

The problem of calculating the positions of the absorption bands in these series is complicated by the existence of many possible *cis-trans* isomerisms and the presence of a group of maxima rather than just one for $n = 3$ or more. For theoretical attack attention is primarily directed to all-*trans* systems and to the principal maximum, with the other maxima regarded as overtones.[39] For isomers that are not all-*trans* there are "*cis*-bands" added to the spectrum that are very characteristic in both location and intensity.

No spectrophotometric problem has received more study than the explanation of the regular bathochromic shifts of the conjugated polyenes. The main relationships proposed for λ_{max} and n are $\lambda^2 = kn$ and $\lambda = kn$, with k a constant in each equation. The first of these was derived from a basic assumption that the chain is a linear oscillator,[40] but empirical or semiempirical modifications give better experimental agreement. Thus, Ferguson cites[41] $\lambda^2 = (0.09 + 2.31n) \, 10^5$ (in angstroms) as a closer approximation, and the best fit of the experimental data appears

[35] F. Bohlmann, *Chem. Ber.* **86**, 63, 657 (1953).

[36] A. E. Gillam and D. H. Hey, *J. Chem. Soc.* **1939**, 1170.

[37] A. Streitwieser, Jr., "Molecular Orbital Theory for Organic Chemists," Chapter 8. Wiley, New York, 1961.

[38] L. N. Ferguson, "Electron Structures of Organic Molecules," p. 284. Prentice-Hall, New York, 1952.

[39] J. Dale, *Acta Chem. Scand.* **11**, 265 (1957).

[40] G. N. Lewis and M. Calvin, *Chem. Revs.* **25**, 273 (1939).

[41] L. N. Ferguson, *Chem. Revs.* **43**, 408 (1948).

to be $\lambda^2 = A - BC^N$, where A, B, and C are constants and N is a function of n.[42] Braude proposed that the length of the conjugated part of the molecule rather than n was proportional to the square of the wavelength.

TABLE II-6

PRINCIPAL ULTRAVIOLET–VISIBLE MAXIMA OF CONJUGATED POLYENES

n	H—(CH=CH)$_n$H[33] Maximum, mμ (ϵ)	CH$_3$—(CH=CH)$_n$—CH$_3$[34] Maximum, mμ (ϵ)	⟨(CH=CH)$_n$⟩ (cyclic)[45] Maximum, mμ (ϵ)
1	162	176 (31,000)	
2	217 (20,000)	227 (51,700)	
3	268 (34,600)	263 (69,200)	
4	304	299 (89,800)	290s (250)
5	334 (121,000)	326 (106,500)	
6	364 (138,000)	352 (124,000)	
7	390		314 (69,000), 374 (5700)
8	410 (108,000)	396 (157,000)	
9		413 (170,200)	269 (7600), 278 (8100), 369 (30,300), 408 (7500), 422 (6800), 448 (21,800)
10	447	437 (192,000)	
12			264 (12,400), 350 (195,000), 364 (201,000), 512 (1740)
15			329 (44,000), 428 (144,000)

While the proportionality of wavelength squared (or cubed for the *p*-polyphenyls[36]) to the number of double bonds is close in the lower members of these series, the higher members show a decreasing increment in wavelength and an apparent convergence, in the polyenes to a value between 600 and 700 mμ. Molecular orbital methods and other sophisticated modern theoretical techniques have been applied to give (among others) the following equations for calculating the wavelength[43] of maximum absorption (or the frequency[44]):

[42] K. Hirayama, *J. Am. Chem. Soc.* **77**, 373 (1955).
[43] W. Kuhn, *Helv. Chim. Acta* **31**, 1780 (1948).
[44] W. T. Simpson, *J. Am. Chem. Soc.* **73**, 5363 (1951); **77**, 6164 (1955).
[45] F. Sondheimer, R. Wolovsky, and Y. Arriel, *J. Am. Chem. Soc.* **84**, 274 (1962).

$$\lambda^2 = \lambda_0^2 \Big/ \left[1 - a \cos\left(\frac{\pi s}{n+1}\right) \right]$$

$$\nu(\text{cm}^{-1}) = \nu_0 - 2\beta \cos \frac{\pi}{n+1}$$

$$\lambda \propto (n+1)^2/(2n+1) \quad (\text{free electron model}[31])$$

In these equations λ_0 and ν_0 are the wavelength or frequency respectively of the maximum for ethylene ($n = 1$), a is a constant, β is the interaction energy, and s is the order of the harmonic (1, 2, etc.).

It is of some interest that the cyanine dyes also contain a conjugated chain of carbon-carbon bonds but both ends of the chain terminate with nitrogens, one of them carrying a positive charge. In this series $\lambda_{\max}$ increases uniformly with the value of n and does not converge at all.

c. Cyclic Polyenes (Nonaromatic).

For the general cyclic structure $\langle(CH{=}CH)_n\rangle$ compounds with n from 4 to 12 are not aromatic. One version of molecular orbital theory predicts a virtually constant $\lambda_{\max}$ in this series rather than the progressive increase with n of the noncyclic polyenes, but other studies disagree.[46] In any case the experimental data (Table II-6) do not at the present time offer any clear trend or obey any obvious rule.

3. NEAR INFRARED

Few data for conjugated polyenes are available, probably because so little of a distinctive nature is to be expected in this region. Butadiene has a strong combination band system near 2.4 μ as well as the expected overtones at lower wavelengths, and conjugated trienes are distinguished by a strong triplet at 2.30, 2.32, and 2.34 μ.[47]

4. INFRARED

The general effect of conjugation on the double bond frequency is a shift to lower values (near 1600 cm^{-1}) than normal, but with an increase in intensity and often a splitting into a number of peaks, the number increasing with the number of conjugated linkages. Conjugation of the double bond with an aromatic ring places the double bond frequency near 1625 cm^{-1}, roughly intermediate between the conjugated and isolated aliphatic double bond frequencies.

[46] M. Gouterman and G. Wagniere, *J. Chem. Phys.* **36**, 1188 (1962).
[47] R. T. Holman and P. R. Edmondson, *Anal. Chem.* **28**, 1533 (1956).

F. Alkynes and Poly-ynes

1. FAR ULTRAVIOLET

Turner[4] states that a terminal triple bond gives a strong maximum below 170 mμ and an inflection near 180 mμ ($\epsilon \sim 2000$), but a nonterminal triple bond has a strong maximum around 175 mμ. Acetylene in the gas phase has strong absorption at 152 mμ and a weaker band at 182 mμ.[48] Compounds of the general type R—C≡CH are characterized by a 186.5 mμ maximum and of the R—C≡CCH$_3$ type by a 190.5 mμ band (both in cyclohexane as solvent).[49]

2. ULTRAVIOLET–VISIBLE

The weak bands in the acetylene spectrum from 210 to 250 mμ[50] contrast sharply with the intense and complex maxima of the conjugated

TABLE II-7

ULTRAVIOLET MAXIMA OF POLYACETYLENES, R—(C≡C)$_n$—R, IN EtOH[51]

n	R	Maxima, mμ (ϵ)
2	H	224, 236, 246
2	Me	218.5(300), 226.5(360), 236(330), 250(160)
3	H	242, 255, 268, 284, 300
3	Me	207(135,000), 239(105), 253(130), 268(200), 286(200), 306(120)
4	Me	205, 215, 226, 234 (all strong) 286, 306, 328, 354 (all weak)
4	Ph	276(133,000), 319(20,000), 331(22,000), 366(31,000) 397(21,000)
5	Me	224, 233.5, 247, 260.5 (all strong) 324, 347.5, 373.5, 394 (all weak)
5[a]	tert-Bu[35]	218(13,000), 228(34,000), 239(125,000), 251(323,000), 265(442,000), 313(158), 339(256), 364(270), 394(156)
6	Me	242(47,000), 255(128,000), 268.5(317,000), 284(445,000)
7[b]	tert-Bu[35]	220(5900), 230(7000), 240(9000), 251(14,000), 263(50,000), 277(160,000), 292.5(395,000), 310.5(527,000), 357.5(700), 384(620), 415(570), 453(290)

[a] In methanol.
[b] In ether.

[48] G. Moe and A. B. F. Duncan, *J. Am. Chem. Soc.* **74**, 3136 (1952).

[49] B. Wojtowiak and R. Romanet, *Compt. rend.* **250**, 2865, 3305 (1960).

[50] F. A. Matsen *in* "Chemical Applications of Spectroscopy" (W. West, ed.), p. 657. Interscience, New York, 1956.

[51] A. E. Gillam and E. S. Stern, "Electronic Absorption Spectroscopy," 2nd ed. Edward Arnold, London, 1957.

poly-ynes which possess some of the highest molar absorptivities (see Table II-7) known in ultraviolet spectrophotometry.[52] The interpretation of the acetylene spectrum has been found extraordinarily complex for so simple a molecule, but the conjugated poly-ynes fit more or less easily into the same class with the polyenes.

In general the conjugated poly-ynes have three band systems, the very strong one at the shortest wavelengths and the second at intermediate wavelengths both exhibiting much vibrational structure. The third band is at long wavelength and too weak ($\epsilon \sim 1$) to be observed[53] in routine spectra.

Conjugation of a double with a triple bond yields an absorption band in the ultraviolet similar in wavelength to that of the corresponding diene but less intense and with a shoulder on the long wavelength side.[51]

3. NEAR INFRARED

The first overtone of the $\equiv$C—H group lies at 1.5–1.6 μ ($\epsilon \sim 1$) with a weak sideband (a "hot" band)[54] about 0.01–0.02 μ higher. In acetylene itself a pair of combination bands near 2.45 and 2.55 μ are of comparable intensity. Since disubstituted acetylenes no longer have a hydrogen attached on the triple bond, no distinctive near infrared absorption should be expected.

4. INFRARED

The $\equiv$C—H stretching fundamental at 3300 cm^{-1} is strong for monosubstituted acetylenes, but of course does not appear at all in disubstituted triple bonds. There is a weak sideband about 20 cm^{-1} lower.[54] The C$\equiv$C stretching vibration lies at 2100–2140 cm^{-1}(s) for monosubstituted acetylenes, but at 2260–2190 cm^{-1} for the disubstituted. The bending vibration of the C$\equiv$C—H is represented by a strong band at 615–642 cm^{-1} in most compounds, though lower (618–578 cm^{-1}) in X—C$\equiv$C—H (X=F, Cl, Br).[55]

In compounds with more than one triple bond the presence of the CH$_2$C$\equiv$C linkage is associated with a strong 1325–1336 cm^{-1} band having an intensity that is proportional to the number of triple bonds in the molecule.[56]

[52] J. B. Armitage, N. Entwistle, E. R. H. Jones, and M. C. Whiting, *J. Chem. Soc.* **1954**, 147.

[53] M. Beer, *J. Chem. Phys.* **25**, 745 (1956).

[54] C. S. Kraihanzel and R. West, *J. Am. Chem. Soc.* **84**, 3670 (1962).

[55] G. R. Hunt and M. K. Wilson, *J. Chem. Phys.* **34**, 1301 (1961).

[56] J. J. Mannion and T. S. Wang, *Spectrochim. Acta* **17**, 990 (1961).

TABLE II-8

CHARACTERISTIC INFRARED BANDS OF ALIPHATIC HYDROCARBON GROUPS[a]

Group	Description of vibration	Frequency (cm^{-1})
C—H		
—(CH$_2$)$_5$—(and up)	Symmetrical bending	1465
	Wagging (crystalline compds.)	1365
	(amorphous-liquid)	1365 and 1350
	Twisting (amorphous)	1305
	Rocking (crystalline)	730 and 720
	(amorphous)	722
—(CH$_2$)$_{1-4}$—	Rocking, for single methylene	770
	for two methylenes	740–750
	for three methylenes	730–740
	for four methylenes	725–730
—CH$_3$	Asymmetrical bending	1450
	Symmetrical bending	1375
—C(CH$_3$)$_2$—	Symmetrical bending	1380 and 1365
—C(CH$_3$)$_3$	Symmetrical bending	1390 and 1365
R$_3$C—H	Bending	1320
C—C		
Methylenes	Stretching	1175 and 1050
—CMe$_2$—	Stretching	1170
—CMe$_3$	Stretching	1250 and 1200
C=C		
Single double bond	Stretching	1660
C=CC=C	Stretching	1600
C=CPh	Stretching	1625
C≡C		
RC≡CH	Stretching	2120
RC≡CR'	Stretching	2220

[a] These values are quoted from reference 22. Stretching frequencies for the C—H group have been omitted (see Table II-1).

Principal infrared frequencies for the aliphatic hydrocarbons so far discussed in this chapter are summarized in Table II-8 except for C—H stretching bands (see Table II-1 for these).

5. FAR INFRARED

No general data exist. Spectra of gaseous HC≡C—X (X=F, Cl, or Br) have been measured up to 40 μ,[55] and a pure rotational spectrum of CH$_3$C≡CH in the 30–50 cm^{-1} region[57] may be of some interest.

[57] D. W. Robinson and D. A. McQuarrie, *J. Chem. Phys.* **32**, 556 (1960).

G. Allenes and Cumulenes

1. FAR ULTRAVIOLET

The data are too thin for general conclusions. In $EtCH{=}C{=}CH_2$ vapor there is a maximum at 181 mμ ($\epsilon \sim 20{,}000$) and a shoulder at 188 mμ ($\epsilon \sim 4000$). The disubstituted allene $MeCH{=}C{=}CHMe$ has maxima at 183 and 194 mμ and a weak absorption near 200 mμ.

2. ULTRAVIOLET–VISIBLE

Many allenes are readily isomerized to acetylenes, and some similarities of allene spectra to those of acetylenes seem to exist. Ethylallene has a weak band at 225 mμ ($\epsilon \sim 500$) in hexane solution. Extended cumulated systems have more intense bands; thus, $CH_2{=}C{=}C{=}CH_2$ has maxima at 241 ($\epsilon \sim 20{,}300$) and 310 mμ ($\epsilon \sim 250$),[58] and the higher cumulenes show a progression of λ_{max} to longer wavelengths with the number of double bonds that resembles that of the conjugated polyynes.[59] Values of λ_{max} and ϵ tend to be higher in the cumulenes than in analogous conjugated polyenes.

3. NEAR INFRARED

Allene has overtone bands at 1.629, 1.16, and 0.85 μ for C—H stretching as well as a combination band at 1.658 μ.[60]

4. INFRARED

The expected $C{=}C$ stretching band at 1600 cm^{-1} is not observed in the allenes, but instead two bands at 1965 and 1070 cm^{-1}, frequencies corresponding roughly to single and triple bond values, are present. The ${=}C{-}H$ frequency, however, is that of a normal olefin.[61]

The infrared spectrum of allene has been determined in all phases, even the crystalline,[62] because of its significance as a model compound for cumulated systems, and the far infrared spectrum to 35 μ has also been reported.[60]

[58] W. M. Schubert, T. H. Liddicote, and W. A. Lanka, *J. Am. Chem. Soc.* **74**, 569 (1952).

[59] F. Bohlmann and K. Kieslich, *Chem. Ber.* **87**, 1363 (1954).

[60] R. C. Lord and P. Venkateswarlu, *J. Chem. Phys.* **20**, 1237 (1952).

[61] A. A. Petrov, T. V. Yakoleva, and V. A. Korner, *Opt. Spectry.* (*USSR*) (*English Transl.*) **7**, 267 (1959).

[62] J. Blanc, C. Brecher and R. S. Halford, *J. Chem. Phys.* **36**, 2654 (1962).

H. Benzene and Substituted Benzenes

There are three principal absorption bands in the electronic spectrum, near 180, 200, and 255 mμ in order of decreasing intensity. The descriptive nomenclature of Doub and Vandenbelt labeled these respectively "second primary," "primary," and "secondary" bands,[63] and the Platt system gives the corresponding electron effects in the notation $1_B \leftarrow 1_A$, $1_{L_a} \leftarrow 1_A$, and $1_{L_b} \leftarrow 1_A$. Other nomenclature will be noted later.

1. Far Ultraviolet

The intense bands in the benzene vapor spectrum at 179 ($\epsilon \sim 47,000$) and 200 mμ ($\epsilon \sim 7000$) (fine structure)[64] appear in solutions at about 185 and 200 mμ. Substituents tend to shift both bands to longer wavelengths

TABLE II-9

Far Ultraviolet Maximum of Some Substituted Benzenes
in Hydrocarbon Solvents[4]

Substituent	$\lambda_{max}(\epsilon)$	Substituent	$\lambda_{max}(\epsilon)$
H	184(50,000)	F	185(58,000)
Me	188(56,000)	Cl	190(55,000)
n-Decyl	190(58,000)	Br	191.5(36,000)
1,2-Me$_2$	192(53,000)	I	196(90,000)
1,3-Me$_2$	194(50,000)	1,2-Cl$_2$	195(32,000)
1,4-Me$_2$	192(54,000)	1,3-Cl$_2$	196(40,000)
1,2,3-Me$_3$	195(55,000)	1,4-Cl$_2$	192(44,000)
1,2,4-Me$_3$	196(51,000)	1,3,5-Cl$_3$	203(50,000)
1,3,5-Me$_3$	199(55,000)	Cl$_6$	216(90,000)
1,2,3,5-Me$_4$	199.5(55,000)	OH	188.6(56,000)
1,2,4,5-Me$_4$	197.5(50,000)	NH$_2$	197(32,000)
tert-Bu	188.7(80,000)	NO$_2$	196(14,000)
C≡CH	200(30,000)	CHO	200(25,000)
CH=CH$_2$	200(20,000)	COOH	196(40,000)
		CN	192(48,000)

and the less intense is thus often found in the ordinary ultraviolet. Substituent effects on the stronger band in the far ultraviolet are indicated in Table II-9.

[63] L. Doub and J. M. Vandenbelt, *J. Am. Chem. Soc.* **69,** 2714 (1947); **71,** 2414 (1948).

[64] J. Grabier, N. Damany-Astoin, and M. Cordier, *Compt. rend.* **251,** 2672 (1960).

2. ULTRAVIOLET–VISIBLE

The low intensity ($\epsilon \sim 200$) band system near 255 mμ in the benzene spectrum is celebrated for its characteristic vibrational resolution even in hydrocarbon solvents, with maxima at 234, 239, 244, 249, 255, 261, and 269 mμ (in frequency units these are evenly spaced harmonics). The entire band is due to a forbidden transition, which accounts for its low intensity and probably also for the fine structure. Since this band has been the most readily accessible, substituent effects on its position and intensity have been studied in much detail (see below). Loss of symmetry through substitution increases the intensity.[65]

An extremely weak and highly forbidden band near 330 mμ in the benzene spectrum has no interest for characterization work because of its weakness.

Compounds containing two or more benzene rings insulated from each other by two or more methylenes or other appropriate groups may be expected to have spectra similar to that of benzene but with a molar absorptivity increased in proportion to the number of benzene rings per molecule. However, benzene rings that are directly linked, as in biphenyl, or joined by conjugation, as in stilbene, have electron interactions between rings that insure spectra quite unlike that of benzene (see Section J in this chapter).

Simply substituted benzenes. The main bands of the benzene spectrum are generally preserved in its substitution products, but there is usually a bathochromic shift, some loss of fine structure and changes of intensity. It is not possible to explore all of the methods of estimating the shifts in position and intensity of the 200 and 255 mμ band systems of benzene produced by substitution because of the sheer volume of literature on this subject.

Most single substituents affect both bands the same; i.e., the ratio of the secondary to the primary band wavelength is about 1.25. The magnitudes of the bathochromic shifts may be correlated with *ortho-para* or *meta* directing characteristics of the substituents or with Hammett or Taft sigma constants. In qualitative terms it has been stated that a conjugative substituent shifts both bands to longer wavelengths, and inductive groups produce much smaller shifts in wavelength but may increase the intensity of the longer wavelength band considerably.[66]

[65] J. Petrushka, *J. Chem. Phys.* **34**, 1120 (1961).

[66] C. N. R. Rao, "Ultraviolet and Visible Spectroscopy," Chapter 5. Butterworths London, 1961.

Of alkyl substituents the methyl group has the greatest effect in increasing both wavelength and intensity. The shift to longer wavelengths of the 200 mμ band produced by a single substituent has been given as follows:[67]

$$NO_2 > CHO > COCH_3 > O^- > COOH = NH_2 > CN = COO^- >$$
$$SO_2NH_2 > OCH_3 > OH > Br > Cl > CH_3 > NH_3{}^+$$

This series (slightly augmented) may be broken into two segments, one for *ortho*- and *para*-directing substituents (electron releasing groups) and the other for *meta*-directing (electron attracting) functions:

o,p-series: $Me_2NH > AcNH > O^- > NH_2 > OCH_3 > OH > Br >$
$$Cl > CH_3$$

m-series: $NO_2 > CHO > COCH_3 > COOH > COO^- = CN >$
$$SO_2NH_2 > NH_3{}^+$$

This division is useful for estimating the wavelengths of maximum absorption in disubstituted benzenes. It has been claimed that p-disubstitution products show a larger bathochromic shift than the sum of the two separate monosubstituted benzene shifts if the two groups are not of the same type (i.e., not both o,p, or m orienting) while groups of the same type effect little greater displacement than that caused by the most displaced monosubstituted compound. In trisubstituted benzenes the maxima lie at wavelengths close to those found in the most displaced monosubstituted benzene of the three possibilities.[68]

Prediction of molar absorptivities in substituted benzenes has also been rather extensively tried. Calculations of the intensities of the 255 mμ (^1L_b) band have, for example, been based on the transition moment of the substituted benzene. In monosubstituted compounds the direction of the moment is taken as the axis through the substituent and bisecting the ring, while in polysubstituted derivatives the transition moment is the vector sum of the moments of the corresponding monosubstituted benzenes (with some qualifications). The intensity is proportional to the square of the vector sum of the individual transition moments for polysubstituted benzenes, or to the square of the moment for monosubstituted ones.

Platt[69] has calculated a table of "spectroscopic moments" for various functional groups to be used to calculate molar absorptivity increments

[67] B. G. Gowenlock and K. J. Morgan, *Spectrochim. Acta* **17**, 310 (1961).

[68] L. Doub and J. M. Vandenbelt, *J. Am. Chem. Soc.* **77**, 4535 (1955).

[69] J. R. Platt, *J. Chem. Phys.* **19**, 263 (1951).

in substituted benzenes (by vector addition in polysubstituted derivatives) as well as condensed ring aromatics and pyridines. In monosubstituted benzenes either large positive or negative values of these moments are accompanied by increased molar absorptivities. Some relations between these moments and wavelengths of the maxima have also been suggested.

The data on which the above generalizations were based are often from rather old sources, and Forbes and co-workers[70] have recently redetermined many benzene derivative spectra in a variety of solvents using more modern instruments.

Some further remarks on ultraviolet spectra of substituted benzenes will be found in later discussions of functional groups in the following chapters.

3. Near Infrared

It appears doubtful that any band in the near infrared spectrum of the average aromatic compound is sufficiently unique to have much value for characterization. The aromatic C—H overtones are at shorter wavelengths than corresponding overtones in saturated hydrocarbons, the first three falling at approximately 1.685, 1.143, and 0.87 μ. In benzene itself combination bands at 2.4–2.5, 1.42, and 1.45 μ, as well as many weaker bands in the 1.35–1.565 μ region have been reported.

Alkylated benzenes are, of course, expected to have methyl and methylene overtones also.

4. Infrared

Aromatic compounds are generally recognized by the aromatic C—H bands near 3030 cm^{-1} and the C=C in-plane vibrations (usually three strong bands) in the 1500–1600 cm^{-1} region.

The aromatic C—H stretching mode is usually in the form of one to three bands that are relatively weak, the most common locations being 3048–3096, 3025–3039, and 3000–3020 cm^{-1} for the three, with the 3030 cm^{-1} location most characteristically observed. The C=C in-plane vibrations may vary widely in intensity with substituents, but one generally lies within 5 cm^{-1} of 1600 and another within 5 cm^{-1} of 1500 cm^{-1}, the third falling in the 1560–1600 cm^{-1} region. The presence of the bands mentioned in this paragraph may be taken as a strong sign of an aromatic compound.

[70] W. F. Forbes, *Can. J. Chem.* **38**, 1104 (1960); **37**, 1294, 1305, 1977 (1959); and earlier papers.

The C—H out-of-plane deformations are all strong bands and their locations are clues to the substitution pattern in the ring: thus, monosubstituted benzenes, which still have five aromatic C—H bonds, have characteristic frequencies of 730–770 and 690–710 cm^{-1}; in compounds with four free adjacent hydrogens (*ortho*-disubstituted benzenes) the characteristic band is at 735–770 cm^{-1}; with three adjacent hydrogens, 750–810 cm^{-1}; two, 800–860 cm^{-1}; and one free hydrogen left on the ring (pentasubstituted benzene), 860–900 cm^{-1}.

Other absorption band groups that show characteristic variations with substitution pattern are the 1650–2000, 650–1000, and 950–1225 cm^{-1} regions. The 650–1000 cm^{-1} bands are mainly the C—H deformations discussed above, and the 1650–2000 cm^{-1} bands are largely summations of these. The distinctive shape of the 1650–2000 cm^{-1} absorption patterns for the various kinds of di-, tri-, and tetrasubstituted derivatives

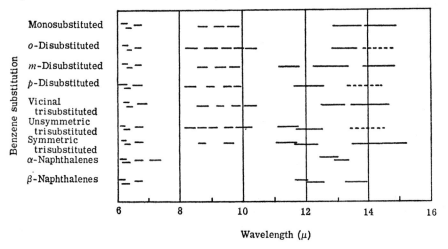

FIG. II-3. Infrared spectra of substituted aromatics, 6–16 μ.

is considered quite helpful for characterization; monosubstituted benzenes, for example, have a series of four main bands in this region decreasing in intensity toward longer wavelengths. Pictures of the usual patterns for different kinds of substitution are available,[22] and Colthup-type charts (see Fig. II-3) may also indicate some of these effects. The 950–1225 cm^{-1} region consists largely of rather weak C—H in-plane vibrations.[10]

Conjugation of a carbonyl or other unsaturated function with the benzene ring generally adds to the spectrum (or enhances) a band near 1580 cm^{-1}.

In a fairly long series of monoalkyl benzenes straight chain substitution accounted for a strong C—H band at 730–750 cm⁻¹ that was shifted to 755–766 cm⁻¹ by branching in the alkyl group.[71]

5. FAR INFRARED

This discussion will be largely limited to alkylbenzenes. Out-of-plane and torsional bending frequencies in the 17–24 μ region give the strongest bands in the cesium bromide region of the spectrum.[11] In many monoalkylbenzenes there is one strong band in the 17.5–22 μ range, a medium one at 16.7–19.2 μ, and weak bands at 14.9–15.3, 24.5–25.0, and 30–35 μ.

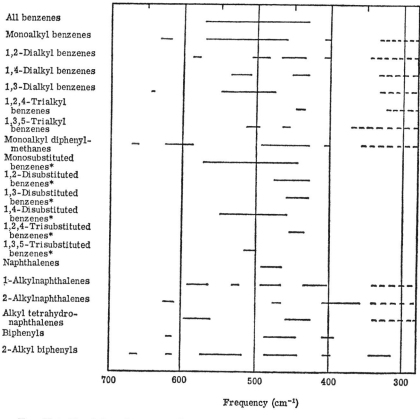

Frequency (cm⁻¹)

Fig. II-4. Far infrared spectra of aromatic hydrocarbons. An asterisk indicates both alkyl and other substituents. Adapted from reference 11.

[71] J. C. Hawkes and A. J. Neale, *Spectrochim. Acta* **16**, 633 (1960).

TABLE II-10
PRINCIPAL FAR INFRARED (15–35 μ) MAXIMA OF SOME
BENZENES AND NAPHTHALENES[a]

Substituents	Maxima (μ)
Benzene	
None	15–16(s), 24.9
Me	16.1, 18.6(w), 19.2, 21.6(s), 29.0
Et	16.0, 18.0(s), 20.6
n-Bu	16.0, 17.2, 17.5(s), 19.7(s), 20.2(s), 22.8
n-Decyl	16.0, 17.0, 17.5(s), 20.3(s)
tert-Bu	16.0, 18.3(s), 21.7, 25.5(w)
1,2-Me$_2$	17.7(s), 20.4(s), 23.5(s), 25.2
1,3-Me$_2$	15.1, 18.6(w), 19.5(s), 23.2(s), 24.9
1,4-Me$_2$	15.5, 20.7(s), 24.4(w), 34.3
1,2,3-Me$_3$	15.3, 18.6(s), 19.5, 20.6, 22.0, 26.0(w)
1,2,4-Me$_3$	18.0, 18.6(s), 21.3, 22.9(s), 31.3
1,2,4,5-Me$_4$	15.0(w), 22.5(s), 34.1(w)
Ph	16.5, 21.8(broad)
Naphthalene	
None	16.2, 21.0(s), 27.8
1-Me	16.1, 17.7, 18.9(s), 19.5, 21.0, 21.5(w), 23.0, 24.5(s), c. 35 (broad)
2-Me	16.1, 21.1(s), 24.6(w)

[a] All data are transcribed from graphs in reference 11.

Dialkyl benzenes show differences that may be correlated with orientation: *o*-dialkyl compounds have a strong band at 21.4–23.0, medium bands at 16.9–17.3, 19.7–20.7, and 29–32, and a weak band at 24.4–24.7 μ; *m*-dialkyl compounds are characterized by a strong absorption at 22.3–23.3 and weak ones in the 18.7–19.7 and 31–35 μ regions; and *p*-dialkyl compounds absorb at 17.5–20.8 with a weak band at 15.5–15.6 μ also. In trialkyl benzenes the following patterns are said to prevail: 1,2,3-trialkyl, 18.7–18.9(s) and 15.3–15.5(m); 1,2,4-trialkyl, 22.7–23.0(s) and 17.0–18.9(s); and 1,3,5-trialkyl, 19.5–20.5(s-m) and 30–35 μ (m, broad). Of the above categories 1,2-dialkyl and 1,2,4-trialkyl compounds have the largest number of bands per compound (see Fig. II-4 and Table II-10).

Optimum wavelengths in the far infrared for analyses of alkylbenzene mixtures have been published.[72] There is an extensive report of far

[72] D. E. Nicholson and S. H. Hastings, *Anal. Chem.* **32**, 138 (1960).

infrared spectra of substituted benzenes,[73] and a few results for halobenzenes from 15 to 40 μ are available.[74]

I. Polynuclear Aromatic Hydrocarbons

1. FAR ULTRAVIOLET

The natural comparison of these compounds with benzene discloses remarkably low intensity absorption by most of the polynuclear hydrocarbons in the 170–200 mμ range; indeed, a number have no maximum

TABLE II-11

FAR ULTRAVIOLET (170–200 mμ) MAXIMA OF SOME POLYNUCLEAR HYDROCARBONS[a]

Compound	λ_{max} (ϵ)	Compound	λ_{max} (ϵ)
Fluorene	None	Anthracene[23]	188.5(21,500)
Naphthalene	195s(10,000)	Acenaphthene	None
1-Methylnaphthalene	None	Naphthacene	185s(17,000)
2-Methylnaphthalene	None	1,2-Benzanthracene	182.5(22,500)
Phenanthrene	188(30,000)	20-Methylcholanthrene	188(25,000)
Chrysene[23]	183(49,000)		
	194(18,000)		
Pyrene	185(55,000)	1,2′-Binaphthyl	196s(25,000)
	197(35,500)		

[a] All data from reference 4 except as shown.

at all here (Table II-11). In many instances this is accounted for by a bathochromic shift of the normal benzene bands by annellation into the ordinary ultraviolet.

2. ULTRAVIOLET–VISIBLE

In a monumental work spanning many years Clar[75] has measured the complex ultraviolet spectra of many condensed ring aromatic hydrocarbons and classified the principal bands.

In hydrocarbon solvents the most immediately striking feature of many of these spectra is pronounced fine structure, often amounting to a dozen or more individual peaks. However, these may usually be classified into three major groups, labeled by Clar β ($\epsilon \sim$ 100,000), p ($\epsilon \sim$ 10,000), and α ($\epsilon \sim$ 100) in order of increasing wavelength. These are

[73] W. S. Wilcox, A. V. Stephenson, and W. C. Coburn, Jr., WAD Division, USAF Pb. 171,300, 1960.

[74] E. K. Plyler, *Discussions Faraday Soc.* **9**, 100 (1950).

[75] E. Clar, "Aromatische Kohlenwasserstoffe," 2nd. ed. Springer, Berlin, 1952.

the equivalents respectively of the 180, 200, and 255 mμ band systems of benzene, and indeed the same nomenclature as for benzene is often used to describe them. In the larger condensed ring systems a very intense β' band may be added on the short wavelength side of these three. A very weak triplet absorption band, ranging from 340 mμ for benzene to 1300 mμ for pentacene, occurs on the long wavelength side.

Molecular orbital theory has had considerable success in correlating the frequencies of these bands with energy differences between orbitals,[37] though the very considerable vibrational structure and overlap of one band system by another offer ample opportunity for confusion. In a recent study[76] of benz[c]anthracene, for example, a spectrum with more than twenty peaks was found to consist of seven distinct but overlapping band systems.

Both α and β bands undergo moderate shifts to longer wavelength with an increase in the number of condensed rings, but the ratio of α to β is nearly constant at about 1.35. The p and triplet bands undergo very large shifts to longer wavelength with linear annellation, but small or even hypsochromic shifts for angular annellation.

In the linear series composed of benzene, naphthalene, anthracene, naphthacene, and pentacene, there is a progressive bathochromic shift with the increase in number of rings (from one to five in the compounds named). The p band shifts so much more rapidly than the others that in pentacene the α band lies between the β and p systems, while in anthracene and naphthacene it is submerged and not observable. In angularly annellated compounds (e.g., phenanthrene, tetraphene, hexaphene) less regular shifts with increasing number of rings occur, and in particular the p band does not show such great alteration.

There are other types of condensed ring aromatics besides the linear and simple angular series described above, and for these it is difficult to offer meaningful description though a thorough mathematical analysis of their spectra is available for some examples. A useful device for characterization is a listing of the longest wavelength maxima of each hydrocarbon (see Table II-12), since these vary over a wide range. Karr[77] has listed the longest wavelength maximum for over 400 polycyclic aromatic hydrocarbons and over 200 polynuclear heterocycles. More detailed accounts of the spectra of naphthalene and derivatives, benzfluorenes, anthracenes, naphthacene, phenanthrene and a few cata- and

[76] J. L. Patenande, P. Sauvageau, and G. Sandorfy, *Spectrochim. Acta* **18**, 257 (1962).

[77] C. Karr, Jr., *Appl. Spectry.* **13**, 15, 40 (1959); **14**, 146 (1960).

TABLE II-12

LONGEST WAVELENGTH MAXIMUM IN THE ULTRAVIOLET–VISIBLE FOR
SELECTED CONDENSED RING HYDROCARBONS[a]

Formula	Name of compound	Solvent	$\lambda_{max}(\log \epsilon)$
$C_{16}H_{10}$	Pyrene	EtOH	334(4.68)
$C_{16}H_{14}$	1,2-Cyclopentenofluorene	EtOH	319(2.38)
$C_{17}H_{12}$	3H-Benzanthrene	Benzene	402(3.54)
	7H-Benzanthrene	EtOH	344(4.17)
$C_{18}H_{12}$	1,2-Benzanthracene	Isooctane	384(2.96)
	Chrysene	EtOH	361(2.80)
	Naphthacene	Benzene	475.5(3.98)
	Triphenylene	EtOH	284(4.23)
$C_{18}H_{14}$	1,2-Dimethylpyrene	EtOH	340(4.47)
	1,6-Dimethylpyrene	EtOH	337(4.63)
	1,7-Dimethylpyrene	EtOH	337(4.60)
	3,5-Dimethylpyrene	EtOH	349(4.59)
$C_{20}H_{12}$	3,4-Benzopyrene	EtOH	403(3.60)
	Perylene	EtOH	434(4.54)
$C_{22}H_{12}$	1,12-Benzperylene	Cyclohexane	385(4.52)
$C_{22}H_{14}$	1,2;3,4-Dibenzanthracene	EtOH	320(3.86)
	1,2;5,6-Dibenzanthracene	Benzene	410(3.1)
	2,3;6,7-Dibenzophenanthrene	EtOH	427(2.76)
	3,4;5,6-Dibenzophenanthrene	EtOH	395(2.38)
$C_{23}H_{16}$	1-Methylpicene	CHCl$_3$	381(2.96)
$C_{24}H_{12}$	Coronene	C$_6$H$_3$Cl$_3$	428(1.95)
$C_{25}H_{16}$	2,3;5,6;8,9-Tribenzoperinaphthene	EtOH	402(3.68)
$C_{26}H_{14}$	Isorubicene	Benzene	625(2.87)
	1,12-o-Phenyleneperylene	Benzene	418(4.47)
	Rubicene	Benzene	530(3.82)
$C_{28}H_{14}$	1,2-Benzocoronene	Benzene	432(2.50)
$C_{28}H_{16}$	2,3;10,11-Dibenzoperylene	Benzene	440(4.54)
	1,2;4,5;8,9-Tribenzopyrane	Benzene	418(2.87)
$C_{30}H_{16}$	Benzo [b] rubicene	Benzene	610(4.26)
	Terylene	Benzene	560(3.6)
$C_{30}H_{18}$	1,2;3,4;5,6;7,8-Tetrabenzoanthracene	C$_6$H$_3$Cl$_3$	382.5(3.24)
$C_{32}H_{16}$	1,2;5,6-Dibenzocoronene	C$_6$H$_3$Cl$_3$	433.5(2.25)
	Naphtho [2,3-a] coronene	Benzene	444.0(2.74)
$C_{34}H_{18}$	4,5;11,12-Dibenzoperopyrene	Benzene	445(4.86)
$C_{36}H_{20}$	1,2;3,4;6,7;12,13-Tetrabenzopentacene	C$_6$H$_3$Cl$_3$	399(3.29)
$C_{42}H_{22}$	4,5;6,7;11,12;13,14-Tetrabenzoperopyrene	Benzene	442(4.34)
$C_{48}H_{22}$	Dicoronyl	C$_6$H$_3$Cl$_3$	432(2.59)

[a] All data are from Vol. IV of "Organic Electronic Spectral Data (1958–1959)."

peri-condensed ring compounds have been summarized in Jaffé and Orchin's book.[31]

3. NEAR INFRARED

The fact that only C—H overtones and combinations are to be expected here makes this region of the spectrum rather uninteresting. Detailed analysis of the aromatic C—H overtones should allow determination of substitution patterns in condensed ring hydrocarbon derivatives, though the infrared is more generally employed for this purpose.

4. INFRARED

It is safe to assume that many of the characteristic features of the benzene spectrum must be preserved in the polycyclic systems in which the spectra also consist of aromatic C—C and C—H vibrations. Thus, bands at 3000–3100, 1500–1600, and near 1490 cm^{-1} as well as strong bands in the 800 cm^{-1} region are to be expected. There are so many different polynuclear hydrocarbons that a wide range of bond strengths and numbers of equivalent hydrogens may be encountered. Some detailed surveys of the spectra of individual ring systems have been published.[78,79]

Analysis of the C—H pattern in the 800 cm^{-1} region is often useful because the number of equivalent hydrogens in a substituted polycyclic compound is a function of the substituent location. For example,[80] naphthalenes with 2-substituents have three sets of frequencies here: the C-1 hydrogen, the pair at C-3 and C-4 and the four equivalent hydrogens on the adjoining ring, absorbing respectively at 835–862, 777–809, and 737–758 cm^{-1}. The first of these varies with the nature of the substituent. In 1-substituted naphthalenes, however, there are only two sets of equivalent hydrogens: the three on the one ring and the four on the other, absorbing respectively at 784–813 and 767–782 cm^{-1}; neither set has much sensitivity to the nature of the 1-substituent. It is then fairly easy to distinguish 1- from 2-substituted naphthalenes. Somewhat similar analyses of 1,2-benzanthracenes[79] and other systems can be performed. One would suspect that nuclear magnetic resonance spectra would do this kind of job with considerably greater ease in general.

[78] C. G. Cannon and G. Sutherland, *Spectrochim. Acta* **4**, 373 (1951).
[79] N. Fuson and M. L. Josien, *J. Am. Chem. Soc.* **78**, 3049 (1956).
[80] T. S. Wang and J. M. Sanders, *Spectrochim. Acta* **15**, 1118 (1959).

5. FAR INFRARED

A few alkylnaphthalenes have been measured in the cesium bromide region, and the 1-alkyl derivatives are said to be richer in absorption bands than the 2-compounds (see Table II-10). Characteristic bands for 2-alkyl naphthalenes are found in the 16.0–16.4 and 21.0–21.5 μ regions, and the 1-alkyl compounds have a distinctive strong band at 23.0–24.9 μ.[11]

J. Ultraviolet–Visible Spectra of Other Hydrocarbons

The substances to be considered here are the azulenes, fulvenes, and other nonalternant hydrocarbons; biphenyls and polyphenyls and the effects of conjugation and steric hindrance on their spectra; stilbenes and related nitrogen analogs; and carbonium ions and carbanions. With the possible exception of the last of these groups it may be expected that spectra in all regions are available, but the ultraviolet–visible spectra are far the most distinctive.

1. NONALTERNANT HYDROCARBONS

Alternant hydrocarbons are those in which the carbons may be divided into two groups (starred and unstarred, if graphic illustration is desired) such that no member of either group is adjacent to a member of the same group; nonalternant hydrocarbons obviously do not have this property and include, in particular, all odd-membered ring hydrocarbons.

If a nonalternant hydrocarbon contains a Hückel number of conjugated atoms ($4j + 2$, where j is an integer), spectra rather similar to the α, β, and p bands of the alternant aromatic hydrocarbons discussed in Section I occur, though some bathochromic shift is usual. In nonalternant systems of $4j$ conjugated atoms this similarity is not so evident, and the absorption pattern may resemble that of a vinyl derivative of the corresponding hydrocarbon with $4j - 2$ conjugated atoms; thus, acenaphthylene has a 1-vinylnaphthalene type of spectrum. Alkyl substituents in nonalternant hydrocarbons do not always have a bathochromic effect as they do in alternant hydrocarbons.

The azulenes are a particularly noteworthy group of nonalternant hydrocarbons in terms of the number of derivatives for which spectra are available and the precision of the interpretation (by Heilbronner and others). As the name implies, the azulenes are generally blue, possessing a series of weak absorption bands near 650 mμ. Plattner's rules, developed empirically and later justified by theory,[81] predict the shift of this visible band system by alkyl substituents. Unlike the alternant hydrocarbons in which alkyl substitution is accompanied by a bathochromic shift, the azulenes exhibit a hypsochromic shift for substituents in the 2,4, and 6 positions and a bathochromic shift in the odd-numbered ring positions. The effects are ten times as great as in alternant hydrocarbons, ranging from 10 to 40 mμ. Polysubstitution of azulenes results in additive frequency shifts. Steric hindrance in the azulenes has been studied in the spectra of acetyl derivatives that contain alkyl groups on adjacent ring positions; a bathochromic shift in the event of hindrance is often observed. This is also a contrast to alternant hydrocarbons in which steric hindrance customarily is associated with hypsochromic changes.

TABLE II-13

PRINCIPAL MAXIMA IN THE ULTRAVIOLET–VISIBLE SPECTRA OF SOME NONALTERNANT HYDROCARBONS[a]

Compound	Maxima, mμ (ϵ)[b]
Fulvene	242(14,000), 373(280)
Azulene	236(4.30), 270(4.79), 340(3.67), 579(2.51), 631.5(2.51)
1-Methylazulene	240(4.30), 280(4.61), 347(3.68), 608(2.42)
2-Methylazulene[c]	610(2.38)
1,3-Dimethylazulene[c]	635(2.69)
4,6,8-Trimethylazulene[c]	547(2.67)
Heptalene[84]	256(21,400), 352(4140)
Acenaphthylene	265(2000), 324(9700), 338(4000)
Pleiadiene	250(25,000), 350(6300), 550(100)
Acepleiadylene	250(20,000), 330(35,000), 550(1500)
1,2;4,5-Dibenzopentalene	281(69,000), 415(15,000)

[a] Except for the azulenes and heptalene data are from reference 83.

[b] Log ϵ is given for the azulenes; data are for hexane solutions and are from Vol. IV of "Organic Electronic Spectral Data." Fine structure bands are omitted.

[c] Partial spectrum only.

The fulvenes are another well studied group of nonalternant compounds, and the spectra of these yellow to red materials have also been analyzed in detail.[82]

[81] E. Heilbronner _in_ "Non-Benzenoid Aromatic Compounds" (D. Ginsburg, ed.). Interscience, New York, 1959.

[82] E. D. Bergmann, _Progr. Org. Chem._ **3**, 81 (1955).

[83] S. F. Mason, _Quart. Rev. (London)_ **15**, 328 (1961).

[84] J. J. Dauben, Jr., and D. J. Bertelli, _J. Am. Chem. Soc._ **83**, 4659 (1961).

Principal maxima in the ultraviolet–visible spectra of a few non-alternant hydrocarbons are listed in Table II-13.[83]

2. BIPHENYLS AND POLYPHENYLS

The ultraviolet spectrum of biphenyl has a broad, intense $\pi \rightarrow \pi^*$ band near 250 mμ that may conceivably be considered a combination of the 200 and 255 mμ bands of benzene. The spectra of 3,3' or 4,4' disubstituted biphenyls are generally similar to that of the parent compound, but the presence of 2,2' substituents of any bulk hinders the

TABLE II-14

ULTRAVIOLET SPECTRA OF SOME BIPHENYLS, POLYPHENYLS, AND STILBENES[a]

Substituents	Maxima, mμ (log ϵ) (solvent)
Biphenyl	
None	251.5(4.26)
4-Methyl	252.3(4.32)(EtOH)
2-Methyl	235(4.02)(EtOH)
2-Ethyl	233(4.05)(EtOH)
2,2'-Dimethyl	227s(3.83), 263.5(2.90), 271(2.78)(EtOH)
Polyphenyls, H—(C$_6$H$_4$)$_n$—H	
$n = 3$, *para*	280(4.40)
meta	251.5(4.64)
$n = 4$, *para*	300(4.59)
$n = 6$, *para*	317.5(4.75)
$n = 10$, *meta*	253(5.33)
$n = 15$, *meta*	254(5.49)
Stilbene	
cis-Stilbene	223(4.28), 274(4.05)(EtOH)
trans-Stilbene	226(4.20), 294(4.45)(EtOH)
2,2'-Dimethylstilbene	230(4.1), 290(4.3)(hexane)
cis-α-Methylstilbene	267 (heptane)
trans-α-Methylstilbene	273.5
cis-α,α'-Dimethylstilbene	252
trans-α,α'-Dimethylstilbene	243.3

[a] Data mostly from Jaffé and Orchin (cf. ref. 31).

normal coplanar configuration of the two rings and the result is the usual hypsochromic shift and decreased molar absorptivity of steric hindrance (see Table II-14). It is well known that sufficiently hindered biphenyls may be resolved into optical isomers, and the ultraviolet spectra offer a rough measure of the possibility of such resolution since resolvable compounds should have spectra closer to benzene than biphenyl in shape. Indeed, with maximum hindrance to coplanarity a biphenyl may have a benzene type spectrum; decamethylbiphenyl, for example, is nearly identical with hexamethylbenzene in the wavelengths of maximum absorption while the molar absorptivities are approximately double (as expected for a molecule with two independent benzene rings).

The binaphthyls are like the biphenyls, but 1,1'-binaphthyl without substituents already has a degree of steric hindrance and a spectrum similar to that of naphthalene. The unhindered 2,2'-binaphthyl has quite a different spectrum.

Polyphenyls in which successive rings are added in the *p*-positions constitute extended conjugated systems for which λ_{max} increases as in the polyenes (see Section E). Only the first few members of the series are available, and for these the cube of λ_{max} is proportional to the number of rings. The corresponding *m*-polyphenyls are not extended conjugated systems and show practically no shift of the principal maximum from the approximately 250 mμ value of biphenyl itself (see Table II-14). The molar absorptivities in the *meta* series increase linearly with the number of rings, indicating that each ring is an independent absorber.

3. STILBENES AND RELATED COMPOUNDS

Stilbenes have *cis-trans* isomers, and the *cis* forms usually absorb at shorter wavelengths and with lesser intensity than the *trans*, as is usually the case for geometric isomer pairs. Suitably placed substituents may exert steric hindrance with resultant hypsochromic shifts also.

Compounds in which one or both double-bond carbons in stilbene are replaced by nitrogen have spectra with much similarity to that of stilbene; such analogs as benzalaniline, azobenzene, azoxybenzene, and phenylnitrone belong in this group (see reference 31 for a detailed comparison).

Some cyclophanes have been prepared in which a double bond links the two benzene rings on one side, and these have spectra corresponding to exceptionally hindered *cis*-stilbenes. Ordinary cyclophanes on the other hand undergo a bathochromic shift relative to open-chain analogs since the steric effect of the cyclophane strain is distributed throughout the molecule.[83] (Here cyclophanes with 2 or 3 methylenes from ring to ring are meant; longer methylene chain bridges suffer no strain and resemble open-chain analogs.)

Styrene may be regarded as a substituted benzene having a large bathochromic shift as a result of the added conjugation.

Substituent effects on biphenyl, *p*-terphenyl and styrene spectra have been studied in some detail.[85]

4. Carbonium Ions and Carbanions

In solutions carbonium ion spectra often result from the solution of an appropriate hydrocarbon or alcohol in concentrated sulfuric acid, and carbanion spectra result from the reaction of an alkali metal with a hydrocarbon in liquid ammonia. In recent years the discovery of the stable tropylium cation and cyclopentadienylide anion and their aromatic properties, both spectroscopic and synthetic, have greatly increased interest in the subject.

Carbonium ions containing conjugated ring systems are of three types: odd-alternant systems such as the triphenylmethyl cation, even-alternant ions, and nonalternant systems such as the tropylium ion.

Both anions and cations from an alternant system, whether even or odd, should have the same long wavelength absorption band, and generally do satisfy this rule at least approximately. The di- and tri- arylmethyl ions have maxima at wavelengths roughly double that of the parent aromatic hydrocarbon.

The tropylium bands at 217 and 274 mμ have been compared to the *p* and *β* bands of ordinary aromatic compounds.

Table II-15 is a condensed guide to the most distinctive regions of the spectrum for various kinds of hydrocarbons.

TABLE II-15

MOST DISTINCTIVE REGIONS OF THE SPECTRUM FOR HYDROCARBONS[a]

Class of hydrocarbon	Best regions	Unsuitable regions
Saturated acyclic	Far infrared, infrared	Ultraviolet–visible
Saturated cyclic		
Cyclopropanes	Near infrared	
Higher rings	Far infrared (?)	Ultraviolet–visible
Monoolefins	Far ultraviolet	
Dienes and polyenes	Ultraviolet–visible	Far ultraviolet
Acetylenes	Infrared	
Poly-ynes	Ultraviolet–visible	
Benzene and derivatives	Ultraviolet–visible, infrared	Near infrared
Polynuclear aromatics	Ultraviolet–visible	
Hydrocarbon ions	Ultraviolet–visible	All other regions

[a] This writer's opinions.

[85] P. Brocklehurst, A. Burawoy, and A. R. Thompson, *Tetrahedron* 10, 102 (1960).

III

Compounds with Oxygen Function

The major groups of compounds to be examined in this chapter are alcohols, aldehydes and ketones and their characterizing derivatives, carboxylic acids together with their esters and other derivatives except amides (see Chapter IV), phenols, and ethers.

Perhaps the most striking feature of the spectra in all regions of most of these functions is the prominent role of hydrogen bonding in determining absorption band positions and intensities and the consequent variations of these with solvent and concentration.

In electronic spectra oxygenated compounds have new possibilities, relative to hydrocarbons, of transitions involving the nonbonding electrons of oxygen to π^* or σ^* states. In vibrational spectra the polar properties of most oxygen functions are responsible for bands considerably more intense than those usually observed in hydrocarbons; not only do the oxygen functions themselves have stronger bands but oxygen enhances the intensity of other vibrations; e.g., the C—C modes that are normally weak in hydrocarbons are stronger in oxygenated derivatives. There are regular decreases in carbonyl stretching frequencies and increases in intensity as bond polarity increases in the order esters, ketones, acids, amides, and acid salts, and considerable success has been enjoyed by attempts to correlate frequency with electronegativity measures such as the Hammett sigma constants.

A. Alcohols

1. FAR ULTRAVIOLET

The simple aliphatic alcohols have an absorption band in the 180–185 mμ region[1] produced by a $n \to \sigma^*$ transition. For methanol vapor λ_{max} is 183.5 mμ ($\epsilon \sim 150$); in hexane solutions a shift of apparent molar absorptivity with concentration indicates association.[2]

---

[1] A. J. Harrison, B. J. Cederholm, and M. A. Terwilliger, *J. Chem. Phys.* **30,** 355 (1959).

[2] W. Kaye and R. Poulson, *Nature* **193,** 675 (1962).

2. Ultraviolet–Visible

Saturated alcohols are fully transparent above 210 mμ and are often used as solvents for the spectra of other compounds when a polar medium is desired. Aromatic and unsaturated alcohols do absorb in the ultraviolet, but the spectra are little different from those of the parent hydrocarbons (see Chapter II).

3. Near Infrared

The fundamental O—H stretching mode for a free hydroxyl lies in the 2.74–2.79 μ region,[3] but association or hydrogen bonding is detected at slightly longer wavelengths that may be more suitably covered by an infrared than a near infrared spectrophotometer.

In dilute carbon tetrachloride solution primary alcohols have the O—H band at 2.750 ± 0.002 ($\epsilon \sim 60$), secondary at 2.760 ± 0.002 ($\epsilon \sim 50$), and tertiary at 2.766 ± 0.003 μ ($\epsilon \sim 45$). Unsaturated and aromatic alcohols often have a shoulder near 2.75 as well as the maximum near 2.766 μ (e.g., benzyl alcohol). Goddu[3] has listed maxima and molar absorptivities for more than fifty alcohols of varied types. All these wavelength assignments assume sufficiently dilute solutions to minimize the customary shift of the hydroxyl band by association or polymerization (see infrared section below for a discussion of this topic).

The first overtone of the O—H stretching mode is near 1.4 μ and has been extensively employed for structure analysis even though the molar absorptivity is rather low ($\epsilon \sim 1$–3). The band is often split through interaction of the hydroxyl with an electron-rich substituent (e.g., double bonds, halogens, aromatic rings) nearby in the molecule. Many alcohols, particularly among the steroids, show doublets for both fundamental and first overtone that may be assigned to conformational pairs,[4] and there are numerous examples of the use of the structure of these bands to make *cis-trans*, syn-anti, and axial-equatorial distinctions in a variety of compounds.

The second overtone is usually at 0.945–0.985 μ in carbon tetrachloride solutions, falling at 0.966 μ for many bicyclic terpene alcohols. This band is also subject to splitting and shift as a function of structure and intramolecular effects. The third overtone for the simple alcohols (except methanol at 0.823 μ) is near 0.738 μ (738 mμ), rising slightly as the homologous series is ascended. This band is extremely weak.

[3] R. F. Goddu, *Advan. Anal. Chem. Instr.* **1,** 347 (1960).

[4] F. Dalton, G. D. Meakins, J. H. Robinson, and W. Zaharia, *J. Chem. Soc.* **1962,** 1566.

Glycols and other polyhydroxy compounds have possible intramolecular hydrogen bonds from one hydroxyl to another that produce distinctive spectra. In the 1,2-glycols free and bonded hydroxyl bands are of roughly equal intensity, falling at about 2.75 and 2.79 μ respectively, but in 1,3-glycols the bonded hydroxyl is only about half as intense as the free one. Though the free hydroxyl has been assigned a frequency of 3636–3643 and the bonded about 3553 cm^{-1}, studies[5,6] of about fifty examples show a correlation of the distance between the two peaks with the C—C—C bond angle. In triols, where two intramolecular hydrogen bonds may exist, the ratio of free to bonded hydroxyl intensities is less than for the diols.[7]

Obviously alcohols will have a full complement of C—H overtones and combinations as well as the O—H frequencies described above. One study of the numerous near infrared maxima in ten normal aliphatic alcohols found the strongest bands at 2.3–2.45 μ, a region consisting mainly of combination bands of various kinds (but wavelengths above 2.6 μ were not included[8]). In a group of simple cyclic alcohols[9] the O—H fundamental and the first two overtones were at 2.74, 1.4, and 0.96 μ respectively and the corresponding C—H bands at 3.4, 1.7, and 1.2 μ. Goddu[3] has surveyed near infrared spectra of alcohols and phenols, and the phenols (see Section E) resemble the alcohols closely (see Fig. II-2 for O—H bands in the near infrared). Much of the data from the overtone regions was compiled in the 1930's.

4. INFRARED

The fundamental hydroxyl stretching mode has been described above, but further discussion of association and polymerization effects tending to shift the band to lower frequencies is presented here.

In polar solvents the sharp, free hydroxyl fundamental at 3590–3650 cm^{-1} is modified to a broad and usually stronger polymer band in the 3200–3400 cm^{-1} region. Dilution of the solution generally shifts the band toward the free hydroxyl frequency because the polymeric structures are formed through intermolecular hydrogen bonding that is weakened

[5] R. Schleyer, *J. Am. Chem. Soc.* **83**, 1370 (1961).

[6] H. Buc and J. Néel, *Compt. rend.* **252**, 1786 (1961).

[7] L. P. Kuhn and R. E. Bowman, *Spectrochim. Acta* **17**, 650 (1961).

[8] R. J. W. LeFèvre, R. Roper, and A. J. Williams, *Australian J. Chem.* **12**, 743 (1959).

[9] O. H. Wheeler and J. L. Matteos, *Bol. Inst. Quim. Univ. Nal. Auton. Mex.* **9**, 22 (1959).

when the concentration decreases and the molecules are separated. In hindered alcohols (e.g., some branched chain aliphatic alcohols) there may be only dimerization and no polymerization through hydrogen bonding, and the hydroxyl frequency is an intermediate value, roughly 3450–3550 cm^{-1}. Intramolecular hydrogen bonding with resonance stabilization, as in the enolic forms of β-diketones, produces a broad, weak hydroxyl band in the 2500–3200 cm^{-1} range and the frequency is not sensitive to concentration changes.

In addition to the O—H stretching bands alcohols have O—H bending vibrations that are generally strong and located as follows: primary alcohols, 1260–1350; secondary, 1260–1350; and tertiary, 1310–1410 cm^{-1}. A C—O stretching band, usually strong, increases about 50 cm^{-1} with each increase in branching; i.e., it is at 1050 for primary, 1100 for secondary, and 1150 cm^{-1} for tertiary alcohols. The presence of α,β-unsaturation or an aryl ring in the α-position lowers these values by about 25 and 40 cm^{-1} respectively. In alicyclic secondary alcohols the frequency decreases with increasing ring size from cyclobutanol (1090 cm^{-1}) to cycloheptanol (1025 cm^{-1}). All alcohols also have broad bands in the 650–750 cm^{-1} range attributed to out-of-plane bonded O—H deformations, but these, like the other bands mentioned in this paragraph, are not exceptionally reliable except as confirmation of structure (see Figs. III-2 and III-3 for summaries of O—H band positions).

Outside of the O—H stretching region the infrared spectra of higher alcohols in the liquid and solid states are not notably different.[10]

5. Far Infrared

A survey of a considerable number of alcohols in the cesium bromide region has confirmed the broad band near 650 cm^{-1} mentioned above. At smaller frequencies torsional and rotational modes have been identified in the simpler alcohols; methanol,[11] for example, has such bands near 190, 160, and 121 cm^{-1}.

B. Aldehydes and Ketones

1. Far Ultraviolet

Both aldehydes and ketones have intense $n \rightarrow \sigma^*$ and $\pi \rightarrow \pi^*$ bands in this region, for acetone vapor at 190 and 166 mμ, respectively. The aldehydes generally have maxima at shorter wavelengths than the

[10] M. Hashikuni, *J. Phys. Soc. Japan* **15**, 941 (1960).
[11] J. K. O'Loane, *J. Chem. Phys.* **21**, 669 (1953).

ketones. The molar absorptivity of the longer band (the only one available without a vacuum spectrograph) of ketones appears to be sensitive to the number of α-hydrogens in the molecule[12] (see Table III-1).

TABLE III-1

FAR ULTRAVIOLET AND ULTRAVIOLET MAXIMA OF SOME SATURATED KETONES RCOR'

R	R'	Far ultraviolet, mμ (ϵ)[12]	Ultraviolet, mμ (ϵ)[13]
Me	Me	186 (1400)	276.5 (12)
			272 (15.5)[a]
Me	Et	185.8 (1100)	277.5 (15)
			273.5 (17.5)[a]
Me	Pr	186.4 (1196)	276 (20)[a]
Me	iso-Pr	186.6 (756)	283.5 (17)
Me	iso-Bu	185.8 (1330)	
Me	tert-Bu	185.5 (1230)	
Me	n-Amyl	185.8 (1160)	
Me	n-Hexyl	186.5 (1204)	
Et	Et		275 (20)[a]
Pr	Bu		280 (23)[a]
Pr	CH$_2$CMe$_3$		287 (28.5)[a]
Me	CH$_2$CH$_2$OH		276 (19)[a]
Pr	CH$_2$CHOHCH$_3$		280 (28)[a]
	Cyclohexanone	186.2s (723)	290 (15.8)
	Cycloheptanone	184s (980)	292 (18.6)

[a] In ethanol solutions[14]; all others in hydrocarbon solvents.

2. ULTRAVIOLET–VISIBLE

The $n \rightarrow \pi^*$ transition is the source of the weak, broad band near 280 mμ ($\epsilon \sim 10$–20) in saturated acyclic ketones and at slightly longer wavelength in aldehydes (HCHO, 310; CH$_3$CHO, about 290 mμ) (see Tables III-1 and III-4). Acetaldehyde and some of its derivatives are relatively unique in having fine structure for this band even in solutions. Cyclic ketones (Table III-2) customarily have the maximum in the 280–300 mμ region, and with the possible exception of cyclobutanone the spectra offer little evidence of ring strain. In the cyclohexanones

[12] D. W. Turner, *in* "Determination of Organic Structures by Physical Methods" (F. C. Nachod and W. D. Phillips, eds.), Volume 2, Chapter 5. Academic Press, New York, 1962.

[13] C. N. R. Rao, G. K. Goldman, and J. Ramachandran, *J. Indian Inst. Sci.* **43**, 10 (1961).

[14] R. Luft, *Ann. chim. (Paris)* **4**, 777 (1959).

conformation affects the wavelength sufficiently to establish axial-equatorial differences in some instances; for example, the equatorial form of an α-bromocyclohexanone has a hypsochromic shift relative to the parent ketone of about 5 mμ while the axial form has a bathochromic shift of about 28 mμ. Similar results have been observed[15] for α-chloro, hydroxy and acetoxy groups. In some 2-arylcyclohexanones the intensity of the band is greater for the axial than the equatorial form.[16]

TABLE III-2

ULTRAVIOLET MAXIMA OF CYCLIC KETONES IN ISOOCTANE SOLUTION[17]

Compound	Maximum, mμ (ϵ)	Compound	Maximum, mμ (ϵ)
Cyclobutanone	281.5 (20)	Cyclopentadecanone	285.9 (23)
Cyclopentanone	300.1 (18)	2-Fluorocyclohexanone	296.5 (18)
Cyclohexanone	291.4 (15)	2-Chlorocyclohexanone	304 (37)
Cycloheptanone	291.8 (17)	2-Chloro-4-*tert*-butyl-cyclohexanone[15]	
Cyclooctanone	290.5 (14)	equatorial	286 (17)
		axial	306 (49)
Cyclodecanone	288.3 (15)		

In α,β-unsaturated carbonyl compounds there is the usual tremendous effect of conjugation, and the spectra are characterized by a high intensity $\pi \rightarrow \pi^*$ band in the 220–260 mμ ($\epsilon \sim 10,000$) range as well as the weak $n \rightarrow \pi^*$ absorption near 320 mμ. The precise wavelength of the intense band is very responsive to substituents in α and β positions, and an empirical summary of these effects has long been available in the form of Woodward's rules and various corollaries. In Table III-3 is given an extended version of these due mostly to Rao.[18]

Cyclopentenones absorb at shorter wavelengths than expected from the table, and a cyclopropenone has been reported to have no ultraviolet maximum at all in this region. Conjugated acetylcyclohexenones with o-substituents show a degree of steric hindrance that is primarily manifested by a sharp decrease of molar absorptivity [acetylcyclohexene, $\lambda_{max} = 232$ mμ ($\epsilon \sim 12,500$); its 2,6,6-trimethyl derivative, $\lambda_{max} = 243$ mμ ($\epsilon \sim 1400$)]; the effect is presumably associated with the assump-

[15] N. L. Allinger, J. Allinger, L. A. Freiberg, R. F. Czaja, and N. A. LeBel, *J. Am. Chem. Soc.* **82**, 5878 (1960).

[16] R. C. Cookson and J. Hudec, *J. Chem. Soc.* **1962**, 429.

[17] E. M. Kosower and G.-S. Wu, *J. Am. Chem. Soc.* **83**, 3144 (1961).

[18] C. N. R. Rao, "Ultra-violet and Visible Spectroscopy." Butterworths, London, 1961.

tion of a *cis*-configuration in the hindered compound with a resulting decrease in effective size of the chromophore. A number of other deviations or exceptions to Woodward's rules, now over twenty years old, have been reported.[19]

TABLE III-3

ESTIMATED PRINCIPAL MAXIMUM IN THE ULTRAVIOLET SPECTRA OF α,β-UNSATURATED CARBONYL COMPOUNDS IN ETHANOL

Ketones	λ_{max} (mμ)	Aldehydes	λ_{max} (mμ)
Unsubstituted	215	Monosubstituted	220
Monosubstituted[a]	225	Disubstituted	230
hydroxyl substituent	250	Trisubstituted	242
Disubstituted			
no exocyclic double bond	235		
one exocyclic double bond	240		
Trisubstituted			
no exocyclic double bond	247		
one exocyclic double bond	252		
two exocyclic double bonds	257		

[a] All substituents are alkyl groups at α and/or β positions, except as shown.

Extension of the conjugated system by another double bond, as in the $\alpha,\beta,\gamma,\delta$-dienones, produces a bathochromic shift of about 30 mμ. Further double bonds produce gradually decreasing increments in wavelength, the series resembling the polyenes in this respect (see Section E of Chapter II).

Although β,γ-unsaturated ketones are not conjugated systems the double bond affects the carbonyl band near 290 mμ by making it extraordinarily strong ($\epsilon \sim 100$–600). In β,γ-benzoketones the band is converted into a rather distinctive quadruplet[20] at 286–318 mμ (isooctane solution).

Conjugation of the benzene ring with the carbonyl results in a spectrum with some typically aromatic characteristics. In benzaldehydes and acetophenones three accessible band systems may usually be recognized: the high intensity first primary at short wavelengths (see Table III-4), the benzenoid secondary band at 270–290 mμ ($\epsilon \sim 1000$–2000) and a weak $n \rightarrow \pi^*$ ($\epsilon \sim 5$–100) band about 320 mμ. Steric hindrance

[19] P. Arnaud and M. Montagne, *Compt. rend.* **251**, 998 (1960).
[20] A. Moscowitz, K. Mislow, M. Glass, and C. Djerassi, *J. Am. Chem. Soc.* **84**, 1945 (1962); K. Mislow *et al.*, ibid., **84**, 1455 (1962).

by *ortho* substituents in acetophenones produces a hypochromic influence on the intense band, an effect also apparent in the benzocyclanones.[21]

Benzophenones may be regarded as partial benzenoid systems in which the basic chromophore is the benzoyl group only slightly modified by the presence of the other benzene ring.[22] In benzophenone itself there is a

TABLE III-4

ULTRAVIOLET–VISIBLE MAXIMA OF ALDEHYDES OF VARIOUS KINDS[a]

Aldehydes	Maxima, mμ (log ε) (solvent)
Saturated aldehydes	
Acetaldehyde	290f(1.2)(cyclohexane)
Chloroacetaldehyde	298f(1.5)(cyclohexane)
Dichloroacetaldehyde	302f(1.6)(cyclohexane)
Chloral	290(1.58)(cyclohexane)
Isobutyraldehyde	290(1.19)(hexane)
Unsaturated aldehydes	
Crotonaldehyde	213(4.24)(cyclohexane)
Tiglaldehyde	222.5(4.22)(cyclohexane)
Senecialdehyde	230(3.81)(hexane)
Cyclohexylideneacetal- dehyde	231.5(4.20)(cyclohexane)
Benzaldehydes (all in hexane solution)	
Pentafluoro	235(4.24), 243(4.17)
m-Chloro	242(4.06), 248(4.00), 288(3.15), 298(3.08)
o-Chloro	246(4.04), 252(3.93), 292(3.24), 300(3.15), 302(3.15)
p-Chloro	253(4.28), 259s(4.19), 276(3.18), 286(3.00)
p-Bromo	257.5(4.27)
p-Fluoro	244(4.11)
o-Methoxy	246(4.02), 253s(3.93), 306(3.66), 314s(3.62)
o-Methyl	243(4.10), 251(4.11), 291(3.23)
o-Amino	255(3.75), 262s(3.54), 352(3.66)

[a] All data from Vol. IV of "Organic Electronic Spectral Data." Interscience, New York, 1963.

single intense band near 250 mμ, but a monosubstituted derivative of general formula $PhCOC_6H_4X$ will have two bands, one for the benzoyl group and the other for the XC_6H_4CO group; with substitution in both

[21] H. H. Jaffé and M. Orchin, "Theory and Applications of Ultraviolet Spectroscopy." Wiley, New York, 1962.

[22] E. J. Moriconi, W. F. O'Connor, and W. F. Forbes, *J. Am. Chem. Soc.* **82**, 5454 (1960); **84**, 3928 (1962).

rings as many as four maxima (though probably overlapping) are to be expected.

Diones in which the second carbonyl is α to the first may be expected to differ in spectra from monocarbonyl compounds because of conjugation. There is the normal $n \rightarrow \pi^*$ weak band near 280 mμ, but also a longer wavelength band (near 450 mμ for glyoxal, $n' \rightarrow \pi^{*'}$, symmetric orbitals involved) in which the wavelength appears to be a function of the angle of twist about the CO—CO bond.[23] Effects of this kind have also been studied in cyclic diketones and in benzils and related compounds.[24] Even β-diketones do not have spectra that can be considered the sum of independent carbonyls because enolization of active hydrogens generally alters the structure.

3. NEAR INFRARED

Many aldehydes have two combination bands near 2.22 and 2.28 μ ($\epsilon \sim 1$); these are at slightly shorter wavelengths for saturated than for α,β-unsaturated aldehydes. Aqueous formaldehyde is unique in having a strong 2.770 μ band,[3] the result of hydration probably.

The fundamental stretching band of the carbonyl in the infrared has overtones at about 2.9, 1.95, 1.45, 1.16, and 0.97 μ. The 2.9 μ band ($\epsilon \sim 2$–4) is too close to the C—H stretching fundamental to be readily observed in most compounds, and the higher overtones are too weak. By comparison with the excellent correlations available for carbonyls in the infrared the near infrared has little to offer.

4. INFRARED

The carbonyl stretching frequency near 1700 cm^{-1} is strong and highly characteristic because of the high sensitivity of both frequency and intensity to structural influence. According to Bellamy,[25] who lists 177 references, the aldehyde and ketone carbonyl band has been the most extensively studied of all infrared correlations.

Aldehydes absorb at slightly greater frequencies than ketones. In saturated aliphatic aldehydes the band is in the 1710–1740 cm^{-1} range, but it is lower for various kinds of unsaturated aldehydes: at 1680–1705 for α,β-unsaturated compounds; 1695–1715 for aromatic aldehydes;

---

[23] N. J. Leonard and P. M. Mader, *J. Am. Chem. Soc.* **72**, 5388 (1950).

[24] N. J. Leonard, A. J. Kresge, and M. Oki, *J. Am. Chem. Soc.* **77**, 5078 (1955).

[25] L. J. Bellamy, "The Infrared Spectra of Complex Molecules," 2nd ed., Chapter 9. Wiley, New York, 1958.

1660–1680 for $\alpha,\beta,\gamma,\delta$-dienals; and 1645–1670 cm^{-1} for β-hydroxy-α,-β-unsaturated aldehydes.

Outside of the carbonyl region further aldehyde characterization is supplied by C—C bands in the 1325–1440 cm^{-1} range for aliphatic aldehydes or in the 1340–1415 cm^{-1} region for aromatic aldehydes; the latter also have bands in the 1260–1320 and 1160–1230 cm^{-1} regions. The C—H stretching frequency of the aldehyde function generally appears as two bands in the 2700–2900 cm^{-1} region, and a C—H deformation of moderate intensity lies at 780–975 cm^{-1}.

Characteristic carbonyl stretching frequencies for ketones of various types are summarized in Table III-5; most of the data are for dilute solutions. Where more than one of the indicated structures is present in the molecule the frequency shifts are usually additive. Though molar absorptivity data are not fully available, they clearly have diagnostic value. In addition to the carbonyl stretchings alkyl ketones have a band in the 1215–1325 cm^{-1} range and diaryl ketones one at 1075–1225 cm^{-1}, but these are not very useful for correlation purposes.

TABLE III-5

INFRARED STRETCHING FREQUENCIES OF THE CARBONYL IN KETONES

Type of compound	Frequency, (cm^{-1})
Saturated acyclic ketones	1705–1725 (c. 180)
7-member ring ketones	1705–1725
6-member ring ketones	1705–1725 (c. 320)
5-member ring ketones	1740–1750
Cyclobutanone	1775
α-Fluoroketones	c. 1770
α-Haloketones (Cl,Br)	1725–1745
α,α'-Dihaloketones (Cl,Br)	1745–1765
Alkyl cyclopropyl ketones	1686–1704
α,β-Unsaturated ketones	1665–1685
Di-α,β-unsaturated ketones	c. 1660
Alkyl aryl ketones	1680–1700
Diaryl ketones	1660–1670 (c. 410)
α-Diketones	1710–1730
β-Diketones	1540–1640
Metal chelate derivatives	1524–1608
γ-Diketones	1705–1725
Quinones	
both carbonyls in one ring	1660–1690
carbonyls in different rings	1635–1655
Tropolones	c. 1600

In both aldehydes and ketones successful correlations of carbonyl stretching frequency with various measures of bond polarity or strength have been reported. A linear relationship between bond order and frequency is one example.[26] In substituted acetophenones of the formula $XC_6H_4COCH_3$ the frequency is low for basic X (e.g., p-OCH_3) and high for acidic (e.g., p-NO_2)[27] and a correlation with the Hammett sigma constants for X exists.[28] Conformation has an influence on the carbonyl band, and solvent effects have been much studied.[29] Trans-annular interactions between sulfur or nitrogen and the carbonyl in medium ring compounds also shift the frequency.[30]

5. FAR INFRARED

In a study of a large number of ketones from 350 to 700 cm^{-1} the most characteristic feature of the spectrum was the large influence of neighboring atoms and groups on the carbonyl.[31] Aliphatic methyl ketones have three principal bands, at 580–600, 510–530, and 385–420 cm^{-1}, and the first of these is also present in aromatic methyl ketones. In aliphatic ketones generally there are two strong bands at 620–630 and 515–540 cm^{-1}, but α-branched compounds have two bands at 565–580 and 550–560 cm^{-1}. Cyclic ketones have a strong absorption in the 480–505 cm^{-1} region, and it is apparent that α-branching and ring formation can be fairly readily detected in the far infrared. In aromatic ketones, however, most of the bands could be attributed to the aromatic ring and no specific correlations for the carbonyl were found. Acetone has bands at 19.1, 20.7, and 25.0 μ; methyl ethyl ketone at 19.45, 21.5, 24.9, and 39.8 μ (the last a methyl torsional mode); and diethyl ketone at 19.4, 22.4, 24.6, and 32.2 μ.

Similar correlations may be expected for aldehydes, though as usual in far infrared spectra, individual differences from compound to compound are more striking than similarities. Acetaldehyde[32] has methyl torsional bands at 150, 262, and 276 cm^{-1}.

[26] G. Berthier, B. Pullman, and J. Pontis, *J. chim. phys.* **49**, 367 (1952).

[27] R. Stewart and K. Yates, *J. Am. Chem. Soc.* **80**, 6355 (1955).

[28] H. H. Freedman, *J. Am. Chem. Soc.* **82**, 2454 (1960).

[29] L. J. Bellamy and R. L. Williams, *Trans. Faraday Soc.* **55**, 14 (1959).

[30] T. W. Milligan and T. L. Brown, *J. Am. Chem. Soc.* **82**, 4075 (1960).

[31] J. E. Katon and F. F. Bentley, *Spectrochim. Acta* **19**, 639 (1963); K. D. Möller, *Compt. rend.* **249**, 2534 (1959).

[32] W. G. Fateley and F. A. Miller, *Spectrochim. Acta* **17**, 857 (1961).

C. Carbonyl Derivatives (Oximes, Semicarbazones, Thiosemicarbazones, and 2,4-Dinitrophenylhydrazones)

These compounds have traditionally been prepared in order to characterize the parent aldehyde or ketone through the melting point of the derivative. Spectrophotometric characterization is often advantageous because the spectra of the derivatives respond more sensitively to minor structural variations than the parent compounds.

1. ULTRAVIOLET–VISIBLE

Oximes have a high intensity absorption at a wavelength little different from that of the parent carbonyl compound (for α,β-unsaturated ketones). The semicarbazones are more distinctive, those of saturated carbonyl compounds having a maximum near 230 mμ ($\epsilon \sim 12{,}000$) and

TABLE III-6

PRINCIPAL ULTRAVIOLET MAXIMUM OF 2,4-DINITROPHENYLHYDRAZONES[a]

Carbonyl compound or type	Maximum (mμ)
HCHO	346
RCHO (R = alkyl)	356–360
α-Halo, α-hydroxy, or α-carbethoxy aldehydes	345–353
RCOCH$_3$	364–366
RCOR'	364–370
C≡C—C=O	363–366
Acrolein	367
C=C—CHO (alkyl substituents)	373–386
C=CCO·R	369–389
Benzaldehydes	370–386
hydroxy or alkoxyl substituents	389–400
ArCOCH$_2$R	377–388[b]
ArCOCHRR' and o-alkylacetophenones	360–388
Benzophenones	385–402
R—(C=C)$_2$—CHO	390–408
R—(C=C)$_2$CO·R'	383–410
R—C=C·CO·C=C—R'	386–400
ArC=C—CO—R (R = H or alkyl)	378–414
R—(C=C)$_3$CHO	400–415

[a] In CHCl$_3$ solution; all data are from reference 34.

[b] Except p-nitro and amino aryl groups.

of α,β-unsaturated aldehydes and ketones at 260–273 mμ ($\epsilon \sim 20,000$). Thiosemicarbazones have the advantage of even higher molar absorptivities; for saturated carbonyl derivatives the maxima are near 229 ($\epsilon \sim 8000$) and 270 mμ ($\epsilon \sim 22,000$) and for α,β-unsaturated compounds at 245 ($\epsilon \sim 10,000$) and 301 mμ ($\epsilon \sim 30,000$).[33]

The 2,4-dinitrophenylhydrazones are the most numerous class of carbonyl derivative, and ultraviolet–visible spectra of nearly a thousand in chloroform solution are available.[34] The principal maximum in the 350–400 mμ vicinity (see Table III-6) responds to changes in the structure of the parent carbonyl compound to a considerable degree. Increased conjugation in the carbonyl compound leads to a bathochromic shift and an increased molar absorptivity. In neutral alcoholic solutions the spectra are not different from those in chloroform, but in alkaline alcohol there are very great changes; for example, saturated ketone derivatives have a maximum near 366 mμ ($\epsilon \sim 23,000$) in ethanol alone, but addition of alkali produces a change to two maxima at 430 ($\epsilon \sim 20,000$–25,000) and 520–535 mμ ($\epsilon \sim 15,000$).

The measured absorbancy of a weighed sample of a 2,4-dinitrophenylhydrazone may be used to calculate the molecular weight of the compound if it can be assumed that only the 2,4-dinitrophenylhydrazone segment of the molecule absorbs at the chosen wavelength.[35]

2. Near Infrared and Infrared

Though infrared spectra of a number of 2,4-dinitrophenylhydrazones[36] as well as of the other carbonyl derivatives have been determined they are of little interest for characterization purposes except for the oximes.

In the near infrared oximes have an O—H fundamental near 2.78 μ ($\epsilon \sim 200$) in carbon tetrachloride solutions that shows an apparent increase in molar absorptivity with decreasing concentration as a result of association of hydroxyls through hydrogen bonding. The first overtone is at about 2.45 μ.

The weak C=N stretching frequency is in the 1653–1684 cm^{-1} (ϵ under 20) region; for acetaldoxime, cyclohexanone oxime and cyclopentanone oxime the respective values are 1675, 1669, and 1684 cm^{-1}.

[33] A. E. Gillam and E. S. Stern, "Electronic Absorption Spectroscopy," p. 118. Edward Arnold, London, 1957.

[34] J. P. Phillips, *J. Org. Chem.* **27**, 1443 (1962).

[35] E. A. Braude and E. R. H. Jones, *J. Chem. Soc.* **1945**, 498.

[36] F. Stitt *et al.*, *Spectrochim. Acta* **17**, 51 (1961).

D. Carboxylic Acids

1. Far Ultraviolet

The higher fatty acids in ethanol solution have a weak maximum ($\epsilon \sim 50$) at 210 mμ in the ultraviolet, but the lower fatty acids absorb about 5 mμ lower. The source of the band is the $n \rightarrow \pi^*$ transition of the carbonyl, tremendously shifted to shorter wavelengths by the hydroxyl.

There is increasing absorption from 185 mμ down to 173 mμ and a maximum below 165 mμ has been observed with vacuum spectrographs.

2. Ultraviolet–Visible

Except for "end absorption" from the band peak at 205–210 mμ the saturated acids are transparent, but conjugation of the carboxyl with double bonds produces $\pi \rightarrow \pi^*$ bands of high intensity; thus, crotonic acid has a maximum near 208 mμ with a molar absorptivity over 10,000. Effects of substituents are similar to those estimated by Woodward's rules for the corresponding ketones: an α,β or β,β disubstitution by alkyl groups increases the maximum to about 217 mμ and trisubstitution (α,β,β) gives a further increase to about 225 mμ.[37] There is a further 5 mμ increment for an exocyclic double bond or for an endocyclic double bond if in a 5- or 7-membered ring. Additional conjugated double bonds or aromatic rings exert the expected large bathochromic effects. A common method for the spectrophotometric analysis of mixtures of oleic, linoleic, and linolenic acids is based on the increased wavelength of maximum absorption of conjugated as compared to unconjugated unsaturated acids.[38]

An aqueous solution of an acid may be partially dissociated to the carboxylate anion unless a little strong mineral acid is added to repress the ionization. If the spectrum of the anion is wanted, the solution may be made alkaline to promote complete dissociation. Except for the saturated aliphatic acids, for which the anion is likely to be as featureless in the ultraviolet as the acid, the spectra of anions differ considerably from those of the undissociated acids. These differences are not only an aid to characterization but also allow the spectrophotometric determination of ionization constants. The lack of color of one of the ionization products, hydrogen ion, and the ease of determining hydrogen ion activity independently with a pH meter are other favorable factors.

[37] A. T. Nielsen, *J. Org. Chem.* **22**, 1539 (1957).

[38] V. C. Mehlenbacher, "Official and Tentative Methods of the American Oil Chemists' Society." 2nd ed., Cd 7-48. American Oil Chemists' Society, Chicago, 1946.

Thus, only the undissociated and dissociated forms of the acid affect the spectrophotometer, and often an appropriate selection of wavelength can make one of these zero. The following paragraphs outline some of the spectrophotometric methods for determining ionization constants; these are very useful, as they apply to phenols and other acidic functions as well as carboxylic acids and also to the conjugate acids of organic bases such as the amines.

If the dissociated form (conjugate base) has no absorbance at the chosen wavelength, the molar absorptivity, ϵ_A, of the pure acid form is determined in a solution of pH sufficiently low to suppress dissociation completely (0.1 N hydrochloric acid is often suitable). Then the absorbance of a solution containing a known amount of total acid in water is measured and the Beer-Lambert law applied to calculate the concentration of acid present in the undissociated form; subtraction of this value from the total acid gives the concentration of the conjugate base and also that of the hydrogen ion. Substitution of these three quantities in the expression for the concentration ionization constant, K_c, gives a numerical value for the latter. Repeated determinations at successively smaller concentrations and extrapolation of the K_c values thus obtained to zero concentration will give a true thermodynamic constant.[39]

Unfortunately this extremely simple method is limited to acids strong enough to be appreciably dissociated in pure water. The following methods give successful results for K_a as low as 10^{-10} or even less, and require only that acid and conjugate base differ in absorption at some wavelength. A wavelength where one form has a maximum absorbance is generally preferred because the small slope of the curve near this wavelength minimizes errors from lack of reproducibility of wavelength settings.

If α is the degree of dissociation of the acid the ionization constant has the form:

$$K_c = c_H \alpha / (1 - \alpha) \tag{1}$$

where c_H is the concentration of hydrogen ion. The absorbance, D_M, of a solution of the acid at a known hydrogen ion concentration such that partial dissociation is achieved is the sum of contributions from the free acid and its conjugate base[40]

$$D_M = (1 - \alpha)D_A + \alpha D_B \tag{2}$$

[39] G. Krotüm, *Z. physik. Chem.* (*B*) **30**, 317 (1935).
[40] W. Stenstrom and N. Goldsmith, *J. Phys. Chem.* **30**, 1683 (1926).

where D_A and D_B are respectively the absorbances of free acid and conjugate base solutions of the same total acid concentration as the mixture. These may often be simply prepared by diluting aliquot portions of a solution of the acid with a buffer, 0.1 N hydrochloric acid and 0.1 N sodium hydroxide respectively for D_M, D_A, and D_B. Substitution of α from Eq. (2) in (1) gives

$$K_c = c_H(D_M - D_A)/(D_B - D_M) \tag{3}$$

High precision from the use of Eq. (3) requires measurements on several solutions of known pH and perhaps a graphical construction.[41-47] The negative logarithm of (3) may be used for the graph:

$$pK_c = pcH + \log (D_B - D_M)/(D_M - D_A)$$

In this equation pK_c is the negative logarithm of the concentration ionization constant and pcH is the concentration pH. A graph of D_M against pcH (see Fig. III-1) has pcH equal to pK_c at the inflection of the curve.

Correction of the ionization constants as calculated above to thermodynamic constants is slightly complicated by the fact that pH as measured with the usual pH meter is considered paH rather than pcH, and the pK value from the above calculation is then not pK_c but a partially corrected constant. For acids of formula HA, with A^- as the conjugate base, the true thermodynamic constant is then:

$$pK = paH + \log (D_B - D_M)/(D_M - D_A) + 0.5\sqrt{u}$$

where u is the ionic strength of the solutions (held constant experimentally for all) and the simple Debye-Huckel equation has been applied. For acids of the form BH^+ with conjugate base B the thermodynamic constant is given by[48]:

$$pK = paH + \log (D_B - D_M)/(D_M - D_A) - 0.5\sqrt{u}$$

[41] R. A. Robinson, M. M. Davis, M. Paabo, and V. E. Bower, *J. Res. Natl. Bur. Standards* **64A,** 347 (1960).

[42] A. Bryson, *J. Am. Chem. Soc.* **82,** 4858 (1960).

[43] V. E. Bowers and R. A. Robinson, *J. Phys. Chem.* **64,** 1078 (1960).

[44] R. A. Robinson and A. I. Biggs, *Trans. Faraday Soc.* **51,** 901 (1955).

[45] R. P. Bell and R. R. Robinson, *Trans. Faraday Soc.* **57,** 965 (1961).

[46] J. P. Phillips, *J. Chem. Educ.* **31,** 81 (1954).

[47] H. Irving, H. S. Rossotti, and G. Harris, *Analyst* **80,** 88 (1955).

[48] A. R. Lawrence and L. N. Ferguson, *J. Org. Chem.* **25,** 1220 (1960).

Although the graphical method described above has been widely used, its shape does not allow maximum utilization of all experimental points and the precise location of the inflection may be difficult. A better procedure constructs a graph of $(D_B - D_M)/(D_M - D_A)$ against c_H; examination of Eq. (3) appropriately rearranged shows that this is a straight line of slope equal to $1/K_c$. If the logarithms of these two functions are plotted against each other the intercept of the resulting straight line is pK_c.[49]

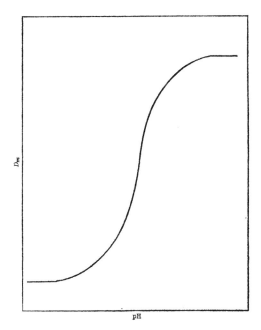

Fig. III-1. Absorbance as a function of pH for acid at a fixed wavelength.

Even if the absorbance of the conjugate base or of the free acid or of both cannot be directly measured appropriate mathematical techniques applied to Eq. (3) allow a graphical solution for the ionization constant. The method can thus be extended to extremely weak acids (for which the pure conjugate base may be unobtainable)[50] and extremely strong

[49] F. J. C. Rossotti and H. Rossotti, "The Determination of Stability Constants, Chapter 13. McGraw-Hill, New York, 1961.

[50] M. T. Beck and M. Halmos, *Nature* **186,** 388 (1960); Stearns and G. Wheland, *J. Am. Chem. Soc.* **69,** 2025 (1947).

ones[51,52] (for which the pure free acid may be unobtainable); straight line plots with slopes related to K_c can be obtained in both instances from rearranged versions of Eq. (3). More elaborate methods requiring solution of three simultaneous equations can also be used.[53,54] If the acid and its conjugate base do not differ appreciably in absorbance, indicator methods are available.[55]

For dibasic acids successive application of the above methods to each ionization separately is possible if the ratio of first to second ionization constant is greater than about 1000, and this is usually the case. For smaller ratios some very elegant, and difficult, spectrophotometric methods have been devised.[56–58] The potentiometric determination is probably superior to the spectrophotometric for these polyfunctional acids.

3. NEAR INFRARED

Since solutions of carboxylic acids in carbon tetrachloride and other inert solvents are equilibrium mixtures of monomeric and dimeric forms, considerable changes in the intensity of the O—H stretching bands in the 2.7–3.0 μ range as a function of concentration are very characteristic. Goddu[3] states that an increase in relative intensity of a 2.82–2.86 μ band upon dilution of a solution 0.02 M or less in initial concentration is a good indication of an acid because other hydroxyl functions vary little in intensity at these low concentrations.

The first overtone of the acid hydroxyl is at about 1.45 μ, and a combination band at 2.1 μ is also characteristic of many acids.

4. INFRARED

A broad, weak band containing a series of minor peaks is found in solid state acid spectra in the 2500–3000 cm^{-1} range. The section from 2500 to 2700 cm^{-1} is considered especially characteristic for the bonded hydroxyl in the carboxylic acid dimer because so few other functional groups absorb in this region at all. In solid state spectra this dimer band

[51] N. Naqvi and Q. Fernando, J. Org. Chem. **25**, 551 (1960).

[52] R. H. Boyd, J. Am. Chem. Soc. **83**, 4288 (1961).

[53] D. H. Rosenblatt, J. Phys. Chem. **58**, 40 (1954).

[54] P. Roman and J. C. Colleter, Compt. rend. **247**, 1456 (1958).

[55] G. Kortüm, W. Vogel, and K. Andrussow, "Dissociation Constants of Organic Acids in Aqueous Solution." Butterworths, London, 1961.

[56] R. A. Robinson and A. I. Biggs, Australian J. Chem. **10**, 128 (1957).

[57] R. A. Robinson and A. K. Kiang, Trans. Faraday Soc. **52**, 327 (1956).

[58] K.-P. Ang, J. Phys. Chem. **62**, 1109 (1958).

may be the only O—H stretching frequency present, but solutions also have the monomeric hydroxyl band (3500–3650 cm⁻¹) described in the near infrared section above. The O—H out-of-plane deformation in acids lies at 900–950 cm⁻¹.

Confirmation of the carboxylic acid function is found in the carbonyl region. The stretching frequency of this function in acids is generally stronger than in ketones and also varies with structural and polar influences. Saturated aliphatic acids in the solid state absorb at 1705–125 cm⁻¹; in dilute solutions in nonpolar solvents separate bands for monomer (higher frequency) and dimer may be present. For some other classes of acids the solid state spectra show the carbonyl band as follows: α-halogen acids, 1720–1740; α,β-unsaturated 1690–1715; aromatic 1680–1700;

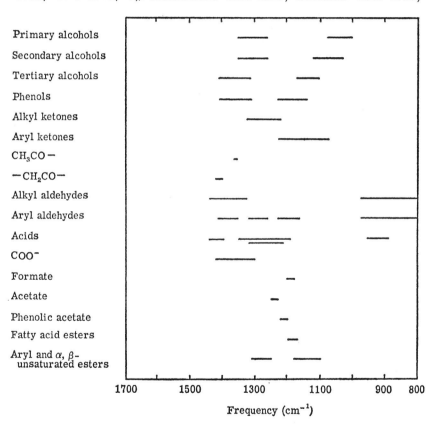

Fig. III-2. Infrared bands of oxygen functions at low frequencies (under 1500 cm⁻¹).

internally hydrogen bonded acids (e.g., *o*-hydroxyaromatic acids), 1650–1670; and dicarboxylic acids, near 1700 cm^{-1}.

The position of the monomeric carbonyl band in substituted benzoic acids has been correlated with substituent basicity;[59] it is at 1720 for the *m*-bromo derivative, 1711 in benzoic acid itself, and 1703 cm^{-1} for *p*-methoxybenzoic acid. Other correlations with Hammett sigma constants, ionization constants and electronegativities have also been proposed.

Acids have a strong band, often two, in the 1210–1320 cm^{-1} region, and a weak absorption near 1400 cm^{-1}; these are associated with a C—O vibration coupled with an O—H in-plane deformation. In the fatty acids there is a series of evenly spaced bands in the 1180–1350 cm^{-1} range that increases in number with increasing chain length; the effect of chain length on band intensities has been studied in some detail.[60]

Infrared spectra of considerably more than a thousand salts of carboxylic acids have been analyzed. Obviously the O—H bands will be missing from these spectra and the carbonyl band replaced by the vibrations of the COO$^-$ group, the asymmetric at 1550–1610 and the symmetric at 1300–1420 cm^{-1}. (Only bands below 1500 cm^{-1} are shown in Fig. III-2.) In doubtful cases conversion of the salt to the free acid and determination of its spectrum as well would be helpful.

5. Far Infrared

Formic and acetic acid vapors have been examined in the 14–50 μ range. Formic acid shows a broad torsional O—H band centered near 40 μ.[61,62] Acetic acid has an O—C—O deformation at 654 cm^{-1}, rocking bands at 582 and 536 cm^{-1}, but no significant bands from 20 to 50 μ.[63,64]

E. Acid Derivatives (Esters, Acid Chlorides, Anhydrides, and Lactones)

1. Ultraviolet–Visible

Unless the alcohol portion of an ester is itself a chromophore, the spectrum in a neutral solvent should not differ appreciably from that

[59] R. Stewart and K. Yates, *J. Am. Chem. Soc.* **82**, 4059 (1960).
[60] R. N. Jones, *Can. J. Chem.* **40**, 321 (1962).
[61] V. Lorenzelli and K. D. Möller, *Compt. rend.* **249**, 520 (1959).
[62] J. K. Wilmshurst, *J. Chem. Phys.* **25**, 478 (1956).
[63] V. Lorenzelli and K. D. Möller, *Compt. rend.* **249**, 669 (1959).
[64] J. K. Wilmshurst, *J. Chem. Phys.* **25**, 1171 (1956).

of the corresponding acid. Saturated esters thus absorb at short wave-lengths [e.g., ethyl acetate at 204 mμ ($\epsilon \sim 60$)]; α,β-unsaturated acid esters show the same bathochromic shifts with α and β substituents as the acids; and aromatic acid esters exhibit modified benzene spectra. Unlike acids, however, spectra of esters in basic solution should not differ from neutral solution, at least until sufficient time for appreciable saponification has elapsed.

Acid anhydrides generally have maxima at slightly longer wavelengths than the parent acid, and acid chlorides are subject to an appreciable bathochromic shift [e.g., acetyl chloride, 235 mμ ($\epsilon \sim 53$)]. Presumably the carbonyl chromophore is less affected by halogen than hydroxyl attached to the same carbon.

2. NEAR INFRARED

Formates have a well defined 2.15 μ band resulting from a combination of C—H with C=O stretching modes.[65] Other types of acid derivatives generally lack unique hydrogen-containing functions and near infrared characterization is thus unlikely to be very helpful. There is a carbonyl overtone near 2.9 μ ($\epsilon \sim 2$–4) for esters in carbon tetrachloride solutions.

3. INFRARED

The carbonyl stretching frequency in esters is usually higher than in ketones or acids, and shows characteristic variations with adjacent structural features (Table III-7). This fact, combined with the positions of the strong C—O bands (see Table III-8), makes the identification of esters not difficult. Lactones may be considered internal esters, but there is some effect of ring strain on the carbonyl frequency in the β and γ lactones.

In acid halides the usual range for the carbonyl band is 1770–1815 cm^{-1}, with conjugated systems near the lower end of this range and saturated compounds at the upper. In acid chlorides there should also be a strong C—Cl band in the 800–600 cm^{-1} region, and for the bromides and iodides analogous bands at still lower frequencies (see Chapter VI). Benzoyl chloride is characterized by a doublet in the carbonyl region at 1779 ($\epsilon \sim 477$) and 1735 cm^{-1} ($\epsilon \sim 177$), and many of its substitution products are similar.[66]

In anhydrides of the acyclic type carbonyl bands are at 1800–1850 and

---

[65] R. M. Powers, M. T. Tetenbaum, and H. Tai, *Anal. Chem.* **34**, 1132 (1962).
[66] C. N. R. Rao and R. Venkataraghavan, *Spectrochim. Acta* **18**, 273 (1962).

TABLE III-7
Infrared Carbonyl Stretching Frequencies of Esters, Lactones Anhydrides, Acid Halides and Peroxides

Compound type	Frequency (cm^{-1})
Saturated aliphatic esters	1735–1750 ($\epsilon \sim 385$)
Esters with α-electronegative substituents	1745–1770
CF_3COOR and $CF_2(NO_2)COOR$[67]	c. 1790
Vinyl esters	1770–1800
α-Keto esters	1740–1755
β-Keto esters (enolic)	1650
γ-(and higher) Keto esters	1735–1750
α,β-Unsaturated and aryl esters	1717–1730
Salicylates and anthranilates	1670–1690
δ-Lactones	1735–1750
γ-Lactones,	
saturated	1760–1780
α,β-unsaturated	1740–1760
β,γ-unsaturated	1800
β-Lactones	1820
Thio esters	1675
Carbonates	1750
Acyclic anhydrides	1740–1780, 1800–1840
Conjugated acyclic anhydrides	1720–1760, 1780–1820
5-Membered ring anhydrides	1760–1800, 1830–1870
Conjugated 5-membered ring anhydrides	1740–1795, 1810–1850
Acid halides	1785–1815
Conjugated acid halides	1770–1800
Acyl peroxides	1780–1800, 1810–1820
Aroyl peroxides	1755–1785, 1780–1805

TABLE III-8
Infrared C—O Stretching Bands of Esters and Acid Anhydrides

Compound type	Frequency, cm^{-1} (ϵ)
Formates	1180–1200 (c. 250)
Acetates	1230–1250 (c. 360)
Phenolic acetates	1200–1220
Propionates and higher esters	1170–1200 (c. 200)
Acrylates, fumarates, maleates	1130–1180, 1200–1300
Benzoates and phthalates	1100–1150, 1250–1310
Acyclic anhydrides	1045–1175
Cyclic anhydrides	1210–1310

[67] R. H. Hughes, R. J. Martin, and N. D. Coggeshall, *J. Phys. Chem.* **24,** 489 (1956).

1740–1790 cm^{-1} and the C—O band is at 1050–1170 cm^{-1}. Conjugation shifts these assignments to smaller frequencies. Anhydrides with 5-membered ring structures have the carbonyl bands at 1820–1870 and 1750–1800 and the C—O in the 1200–1300 cm^{-1} range. (Some disagreement among various sources is shown in the slightly different ranges for these functions of Table III-7 and Table III-8.)

Some help in identifying acid derivatives is provided by analysis of frequencies due to derivative components; thus, bands due to alkyl groups in various kinds of esters have been studied in some detail.[68]

4. Far Infrared

Methyl formate has an O—C—O deformation at 638 cm^{-1} and a C—O—C deformation at 327 cm^{-1}, and methyl acetate has corresponding bands at 646 and 305 cm^{-1} for the vapor phase spectrum[63] and also carboxy rocking bands at 613 and 433 cm^{-1}.

F. Phenols

1. Ultraviolet–Visible

The spectrum of a phenol in inert solvents may be considered a modified benzene spectrum, as the normal benzene bands are present though at increased wavelength and intensity due to the polar nature of the hydroxyl group. Intermolecular hydrogen bonding is evidenced by a decrease in wavelength of the two principal bands (near 210 and 270 mμ in phenol) with increasing concentration; hydrogen bonding is not appreciable below 0.01 M concentrations, however.

Phenols are weak acids (K_a about 10^{-10}), and the formation of the phenoxide ion in basic solutions is accompanied by bathochromic and hyperchromic shifts of considerable magnitude (see Table III-9). The determination of ionization constants of phenols by the spectrophotometric methods described earlier for acids is especially valuable because these constants are too small for potentiometric measurement in most cases.

As derivatives of phenols the indophenols are particularly interesting because of their characteristic intense color; a large number of spectra of these have been reported.[69]

[68] A. R. Katritzky, J. M. Lagowski, and J. A. T. Beard, *Spectrochim. Acta* **16**, 954, 964 (1960); A. R. Katritzky *et al.*, *J. Chem. Soc.* **1958**, 2182.

[69] D. N. Kramer, R. M. Gamson, and F. M. Miller, *J. Org. Chem.* **24**, 1742 (1959).

TABLE III-9

ULTRAVIOLET MAXIMA OF SOME SIMPLE PHENOLS[a]

Compound	Maxima, mμ (log ϵ) (solvent)
m-Cresol	214(3.79), 271(3.20), 277(3.27) (cyclohexane)
	238(3.91), 289(3.51) (0.01N NaOH)
o-Cresol	213(3.85), 270(3.27), 276(3.24) (cyclohexane)
	235(3.95), 286(3.51) (N NaOH)
p-Cresol	236(3.89), 295(3.50) (0.01N NaOH)
2,3-Dimethylphenol	269(3.14), 273(3.15), 278(3.19) (hexane)
2,6-Dimethylphenol	212(3.90), 269(3.18), 275(3.20) (cyclohexane)
2-Ethylphenol	212(3.87), 263s(3.15), 270(3.30),
	276(3.28) (cyclohexane)
o-Nitrophenol	270(3.87), 342(3.58) (cyclohexane)
p-Nitrophenol	295(4.04) (cyclohexane)
o-Fluorophenol	283 (3.43) (0.1N NaOH)
p-Fluorophenol	298 (3.47) (0.1N NaOH)
o-Bromophenol	295 (3.60) (0.1N NaOH)
p-Bromophenol	298 (3.36) (0.1N NaOH)
1-Naphthol	327s(3.90), 332(3.93) 225(4.5), (0.2N NaOH)
	285(3.5), 295(3.7), 325(3.5) (cyclohexane)
2-Naphthol	272s(3.68), 282(3.79), 292(3.68),
	346(3.47) (0.2N NaOH)

[a] All data from Vol. IV of "Organic Electronic Spectral Data."

2. NEAR INFRARED

The sharp free hydroxyl band at about 2.8 μ ($\epsilon \sim$ 170–200) is more intense than in alcohols. Goddu[3] gives the wavelength as 2.772 ± 0.003 μ for most phenols. Bulky 2,6-substituents shift the band to shorter wavelengths and compounds in which there is an intramolecular influence from an *o*-substituent (e.g., *o*-phenylphenol) other than steric hindrance often absorb at longer wavelengths. The intensity of the O—H band decreases with increasing temperature.[67]

The first overtone region near 1.40 μ was intensively studied in the 1930's in connection with the effect of *ortho* substitution. In the *o*-halophenols, for example, separate bands for *cis* and *trans* positions of hydroxyl hydrogen with respect to halogen were identified. In general, phenols with amino, ether, thioether, allyl, halogen, or hydroxyl groups in the *ortho* position may be expected to have an intramolecularly bonded hydroxyl band at slightly longer wavelength than the free hydroxyl band and in addition to it. Since *meta* and *para* substituents cannot produce these effects, *ortho* substitution is identifiable. There is a large literature[3] on these phenomena in both the fundamental and first overtone regions, and similar results have also been studied in the second overtone.

Phenols also have a weak combination band near 2.07 μ.

3. INFRARED

The hydroxyl stretching region has been discussed in the near infrared section above (see also Fig. III-3). A strong O—H deformation frequency and a C—O stretching band at 1200 and 1310–1410 cm^{-1} (Fig. III-2)

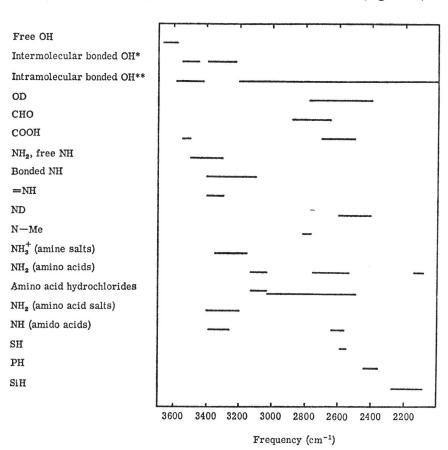

Free OH

Intermolecular bonded OH*

Intramolecular bonded OH**

OD

CHO

COOH

NH$_2$, free NH

Bonded NH

=NH

ND

N—Me

NH$_3^+$ (amine salts)

NH$_2$ (amino acids)

Amino acid hydrochlorides

NH$_2$ (amino acid salts)

NH (amido acids)

SH

PH

SiH

3600 3400 3200 3000 2800 2600 2400 2200

Frequency (cm^{-1})

FIG. III-3. Hydrogen stretching bands, mainly for oxygen and nitrogen functions, in the infrared. A single asterisk indicates that the left band is for dimers, the right for polymers. A double asterisk indicates the broad band is for chelates.

are noteworthy infrared characteristics. Shrewsbury[70] has determined the spectra of 111 alkyl phenols in carbon disulfide from 650 to 1400 cm^{-1}, and gives schematic spectra for all, as well as charts of characteristic frequency patterns for substituent positions.

4. Far Infrared

Spectra of phenol, resorcinol, hydroquinone and pyrocatechol have been determined to 40 μ under various conditions.[71] In Nujol there are bands for phenol at 683, 615, 526, 505, 452, 418, 413, and 265 cm^{-1}; for resorcinol at 678, 656, 616, 543, 494, 460, 386, 356, and 256; for hydroquinone at 677, 610, 544, 517, 457, 410, and 387; and for pyrocatechol at 626, 560, 548, 498, 448, 375, 334, and 277 cm^{-1}.

G. Ethers and Peroxides

1. Far Ultraviolet

In the vapor phase dimethyl ether has a maximum at 184 mμ ($\epsilon \sim 2520$), and a band in the 188–190 mμ interval has been noted[72] for a number of other aliphatic and cyclic ethers. Turner[12] has stated that an absorption band for ethers lies near 170 mμ, and both bands have been identified in the cyclic ethers tetrahydrofuran and tetrahydropyran.

2. Ultraviolet–Visible

Except for some possible end absorption the aliphatic ethers are transparent. Conjugation of the epoxy group with carbonyl in suitable compounds has been detected through a bathochromic shift as the result of the interaction of electrons of the two functions.[73]

Aromatic alkyl ethers have benzenoid spectra (see Chapter II) not much different from the corresponding phenols (except that no change is expected when a neutral solution is made basic). Anisole thus resembles phenol, but a bulky substituent in the *ortho* position exerts sufficient steric effect to be detected by a decrease in the 270 mμ band intensity. In diaryl ethers with large substituents adjacent to the oxygen there

---

[70] D. D. Shrewsbury, *Spectrochim. Acta* **16**, 1298 (1960).

[71] A. Hidalgo and C. Otero, *Spectrochim. Acta* **16**, 528 (1960).

[72] A. J. Harrison and D. R. W. Price, *J. Chem. Phys.* **30**, 357 (1959).

[73] N. H. Cromwell, F. H. Schumacher, and J. L. Adelfang, *J. Am. Chem. Soc.* **83**, 974 (1961).

can be a much greater steric hindrance and this maximum may be eliminated totally.[74]

3. NEAR INFRARED

Terminal epoxides,

$$\begin{array}{c} \text{O} \\ / \quad \backslash \\ \text{—CH——CH}_2, \end{array}$$

have a very characteristic strong 2.20 μ combination band and a weaker ($\epsilon \sim 0.2$) first overtone of the C—H at 1.65 μ. The 2.20 μ band is not subject to interference by bands of other kinds of ethers.

The near infrared has also been recommended as a region of choice characterization for hydroperoxides. These have hydroxyl bands at about 2.81–2.84 μ ($\epsilon \sim 30$–90), and this absorption appears as a doublet for many aralkyl representatives. There are also useful bands near 2.09 μ (ϵ near 1, but higher for aromatic than aliphatic hydroperoxides) and 1.46 μ. The latter is generally a doublet for aralkyl compounds, each component having about half the intensity of the single band in aliphatic hydroperoxides.

4. INFRARED

The only important feature of the spectrum of an ether here is the strong C—O stretching band, but as this is also possessed by alcohols and other compounds its value is limited. Bellamy lists the following frequencies for various kinds of ethers: alkyl (—CH$_2$OCH$_2$—), 1060–1150; aryl and other $=$C—O function ethers, 1230–1270; epoxy compounds, 1250, and also a *cis* band near 830 or a *trans* band near 890; and higher cyclic ethers, 1070–1140 cm^{-1}. Cross[75] also lists bands for trimethylene oxides at 970–980; for phthalans at 895–915; for OCH$_2$O at 940; and for *tert*-butoxy groups at 800–920 cm^{-1}.

The C—H stretching vibrations in ethers are medium or weak. Cross gives some characteristic values (see also Table II-1): OMe, 2815–2830; epoxy, 2990–3050; alkyl acetals, about 2825; OCH$_2$O, about 2780; and C$=$CHO—, 3050–3150 cm^{-1}.

Alkyl peroxides have weak O—O bands in the 830–890 cm^{-1} region; these are near 1000 cm^{-1} in aryl peroxides.

[74] M. Dahlgard and R. Q. Brewster, *J. Am. Chem. Soc.* **80,** 5861 (1958).
[75] A. D. Cross, "Practical Infra-Red Spectroscopy." Butterworths, London, 1960.

5. Far Infrared

Aliphatic ethers generally have an absorption band in the 480–520 cm^{-1} (20 μ) region. Trimethylene oxide has bands attributed for the most part to puckering vibrations at 250–300 cm^{-1} (broad, weak) in the vapor phase, and in the liquid phase a weak band at 400 cm^{-1}.[76]

Table III-10 gives a highly abbreviated summary of the principal conclusions to be drawn from this chapter.

TABLE III-10

Most Distinctive Regions of the Spectrum for Oxygen-Containing Groups[a]

Type of compound	Best region(s)	Outstanding feature
Alcohols	Near infrared, infrared	Shifts due to hydrogen bonding
Glycols	Near infrared	Hydrogen bonding
Aldehydes and ketones	Infrared	Carbonyl stretching
Diones	Ultraviolet	Angle of twist
$\alpha,-\beta$Unsaturated	Ultraviolet	Woodward's rules
Carbonyl derivatives		
Oximes	Near infrared	
Semicarbazones	Ultraviolet	
2,4-Dinitrophenylhydrazones	Ultraviolet	
Carboxylic acids	Ultraviolet	Changes with pH
	Infrared	Association
Acid derivatives		
Formates	Near infrared	
All esters	Infrared	Carbonyl variation with structure
Acid halides	Infrared	
Anhydrides	Infrared	
Phenols	Ultraviolet	Changes at high pH
	Near infrared	Steric effects
Ethers	All mediocre	
Hydroperoxides	Near infrared	

[a] These are opinions.

[76] A. Danti, W. J. Lafferty, and R. C. Lord, *J. Chem. Phys.* **33**, 294 (1960); S. Chan, J. Zinn, and W. D. Gwinn, *J. Chem. Phys.* **33**, 295 (1960).

IV

Compounds with Nitrogen Functions

Only the simpler nitrogen functions are considered in this chapter; some nitrogen heterocycles are considered in Chapter V and amino acids, peptides, proteins, purines, nucleic acids, and other complex substances with biochemical significance in Chapter VIII.

In their electronic spectra nitrogen compounds offer possible $n \rightarrow \sigma^*$ and $n \rightarrow \pi^*$ transitions due to the free pair of electrons on the nitrogen. These electrons are often used in bonding to hydrogen or other elements and this basic characteristic of many nitrogen functions accounts for changes in spectra in going from neutral to acid solutions. The disappearance of an absorption band for a nitrogen base in passing from an inert to an acidic solvent is said to be the sole solvent effect that can be attributed with certainty to the loss of a transition involving lone pair electrons.

In vibrational spectra hydrogen bonding effects are important factors as with oxygen compounds. In general, the behavior of nitrogen in organic compounds is more varied and complex than that of oxygen and the interpretation of infrared spectra correspondingly less exact. The close correlation of absorption bands with particular functions is less precise in nitrogen compounds; for example, in amide spectra there are some bands without definite functional assignment and designated by number only.

A. Amines

1. Far Ultraviolet and Ultraviolet–Visible

Saturated aliphatic amines have a lone pair of electrons responsible for a rather weak ($\epsilon < 1000$)[1] $n \rightarrow \sigma^*$ band near the lower edge of the normal ultraviolet (MeNH$_2$, 215; Me$_2$NH, 220; and Me$_3$N, 227 mμ). This band is lost in acid solutions because the electrons are then employed in a nitrogen-hydrogen bond. In the far ultraviolet the saturated amines have a stronger ($\epsilon \sim 2000$–5000) band (MeNH$_2$, 173.7; Me$_2$NH, 190.5; and Me$_3$N, 199 mμ).[2]

[1] S. F. Mason, *Quart. Revs.* (*London*) **15**, 287 (1961).
[2] E. Tannenbaum, E. Coffin, and A. Harrison, *J. Chem. Phys.* **21**, 311 (1953).

The amino group in compounds containing interacting additional functions usually exerts an auxochromic effect, and a dialkylamino group in particular can produce a very considerable bathochromic shift. The simple aromatic amines have spectra similar to those of the parent aromatic hydrocarbon but with a considerable bathochromic shift, an increase in intensity and a loss of vibrational structure. For example, the 200 mμ band of the benzene spectrum is at 230 mμ in aniline, and the 255 mμ system is replaced by a single stronger maximum near 280 mμ; however, in acid solution the spectrum reverts very nearly to that of benzene, with main bands at 203 and 254 mμ, since the use of the free nitrogen electrons of aniline in the bond to hydrogen effectively eliminates the auxochromic action of the amino group.

TABLE IV-1

STERIC HINDRANCE EFFECTS IN THE ULTRAVIOLET SPECTRA OF ANILINES[a]

Substituents in aniline	Maxima, mμ (ϵ)
N-Me	243(13,200), 295(2300)
N-Me-2,4,6-(tert-Bu)$_3$	247(4900)
N-Hexadecyl	245(14,200), 295(2250)
N,N-Me$_2$	176, 200, 250, 296
N,N,4-Me$_3$	183, 203, 254, 304
N,N,2-Me$_3$	185, 207, 248
N,N,2,6-Me$_4$	194, 209, 259
N,N-Me$_2$[b]	251(15,500)
N,N-Me$_2$-2-Et[b]	249(4950)
N,N-Me$_2$-2-iso-Pr[b]	248(4300)
N,N-Me$_2$-2-tert-Bu[b]	Lacking
N,N,2,4,6-Me$_5$[b]	257(2500)
4-NO$_2$	376(15,500)[c]
4-NO$_2$-3-Me	374(13,200)[c]
4-NO$_2$-2,3-Me$_2$	382(9750)[c]
4-NO$_2$-3,5-Me$_2$	385(4840)[c]
4-NO$_2$-2,3,5,6-Me$_4$	396(1560)[c]
4-NO$_2$-3,5-(tert-Bu)$_2$	401(540)[c]

[a] Most of these data are for hydrocarbon solvents and are from reference 3.

[b] Partial spectrum only.

[c] Partial spectrum in ethanol.

The principal maxima in N,N-dimethylaniline are at 180, 210, 250, and 300 mμ, the last having no counterpart in the benzene spectrum. Bulky ortho substituents produce steric hindrance to coplanarity of the —NMe$_2$ group with the ring and a large decrease in the molar absorptivity of the 250 mμ band and total disappearance of the 300 mμ band may result. The angle of twist out of the plane can be computed from the magnitude of the decrease in molar absorptivity (see Table IV-1 for illustrative spectra).[3] The most highly hindered derivatives have spectra similar to those of alkyl benzenes.

Steric effects in *p*-nitroanilines have been subjected to detailed analysis also;[3] here hindrance is accompanied by a bathochromic shift and a decrease in intensity (see Table IV-1). Steric effects in ultraviolet spectra cannot be covered in this book, but it is worth noting that most such effects produce decreases in λ_{max} or ϵ_{max} or both.

2. NEAR INFRARED

The fundamental N—H stretching mode at 2.8–3.0 μ and its first and second overtones at 1.5 and 1.0 μ all serve to distinguish primary and secondary amines from tertiary, since the latter obviously must lack N—H bands altogether (see Fig. IV-1).

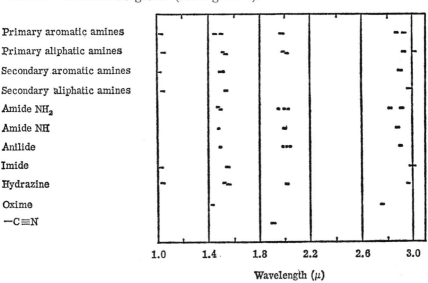

Primary aromatic amines
Primary aliphatic amines
Secondary aromatic amines
Secondary aliphatic amines
Amide NH$_2$
Amide NH
Anilide
Imide
Hydrazine
Oxime
—C≡N

1.0   1.4   1.8   2.2   2.6   3.0

Wavelength (μ)

FIG. IV-1. Near infrared bands of nitrogen functions. Adapted from reference 4.

Primary amines generally have separate symmetric and asymmetric peaks for the N—H stretching (see infrared section following), with the

[3] H. H. Jaffé and M. Orchin, "Theory and Applications of Ultraviolet Spectroscopy," Wiley, New York, 1962.

[4] R. F. Goddu, *Advan. Anal. Chem. Instr.* 1, 347 (1960).

molar absorptivities of aromatic amines ($\epsilon \sim 20$–60) about twenty times greater than those of aliphatic. There are also a combination band near 2.0 μ, a doublet in the 1.45–1.55 μ overtone region and a single weak 1.0 μ peak. In about 40 spectra of primary aromatic amines in carbon tetrachloride the combination band ranged from 1.959 to 1.982 μ ($\epsilon \sim 1.75$), the symmetric overtone was near 1.492 μ ($\epsilon \sim 1.4$) and the weaker asymmetric band near 1.447 μ.[4] In substituted anilines the position of the first overtone varies with the nature of the substituents,[5] and correlation of frequency of fundamental and overtone with Hammett sigma factors can be demonstrated.

Secondary amines have only a single peak in both the fundamental 2.9–3.0 μ region and the 1.50–1.55 μ overtone because there is only one N—H bond in the functional group rather than two as in the primary amines. Since secondary amines also lack the 2.0 μ combination peak, the difference of primary and secondary amines in the near infrared is rather pronounced.

3. INFRARED

Though the N—H fundamental was described in terms of wavelength in the near infrared section above, it is repeated here in frequency units: primary amines have a symmetric and asymmetric pair related by the empirical equation $\nu_{sym} = 345.5 + 0.876\nu_{asym}$ in the 3400–3500 cm^{-1} region but secondary amines have only a single weak band in the 3310–3350 cm^{-1} range (see also Fig. III-3). Hydrogen bonding may lower these frequencies, but not to as great an extent as for oxygen compounds, and solid state spectra also tend to yield lower frequency assignments than solutions. Aralkyl secondary amines absorb at about 3450 cm^{-1}, approximately 100 cm^{-1} higher than dialkyl amines. In anilines the asymmetric frequency is more intense for *ortho* substituted derivatives than for *meta* or *para*, and the frequency can be correlated with the twist of the H—N—H angle by steric interaction.[6]

In-plane and out-of-plane bending vibrations for primary amines are respectively 1560–1640 and 650–900 cm^{-1}. In secondary amines a weak 1490–1580 cm^{-1} band represents N—H bending.[7]

The C—N stretching vibrations (Table IV-2) do not differ much in position from C—C stretching bands but are strong because of the

[5] K. B. Whetsel, *Spectrochim. Acta* **17**, 614 (1961).

[6] P. J. Krueger, *Can. J. Chem.* **40**, 2300 (1962).

[7] K. Nakanishi, "Infrared Absorption Spectroscopy," Holden-Day, San Francisco, 1962.

greater polarity of the C—N linkage. In aryl or mixed aryl-alkyl amines these are found at 1250–1360 cm^{-1} for the aromatic and 1180–1280 cm^{-1} for the aliphatic C—N linkage. Medium bands in the 1030–1230 cm^{-1} region characterize aliphatic amines. See also Fig. IV-2.

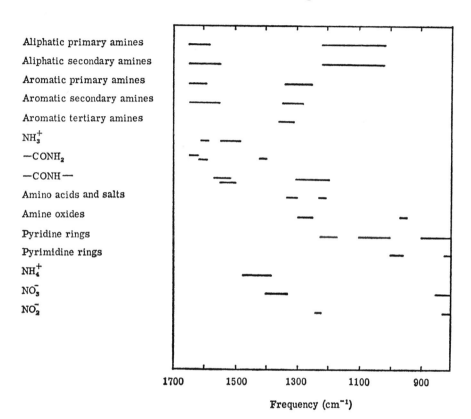

FIG. IV-2. Low frequency vibrations in the infrared for nitrogen functions. Adapted from A. D. Cross, "Introduction to Practical Infra-Red Spectroscopy." Butterworths, London, 1960.

Amine hydrochlorides and other derivatives containing the —NH$_3^+$, —NH$_2^+$ or —NH$^+$ linkages show stretching and deformation bands as indicated in Table IV-2 and (perhaps rather confusingly) Fig. III-3. Possibly the best way to identify an amine by infrared spectroscopy is to treat the sample with acid and look for the broad ammonium band in the 2200–3000 cm^{-1} region.

TABLE IV-2
INFRARED STRETCHING AND BENDING FREQUENCIES OF AMINES AND SALTS

Infrared stretching and bending	Frequency (cm^{-1})
N—H Stretching	
Primary amines	3500 and 3400
Dialkyl amines	3310–3350 (w)
Aryl alkyl amines	3450
—NH_3^+	3000 (broad)
	2500 and 2000 (overtones)
—NH_2^+ and —NH^+	2250–2700
$\overset{+}{C}=N—H$	2300–2500
Guanidinium	1800–2200 (m)
	3300
N—H Bending	
—NH_2	1560–1640 (in-plane)
	650–900 (out-of-plane)
—NH—	1490–1580 (w)
—NH_3^+	1500 and 1575–1600
—NH_2^+—	1575–1600 (m)
$\overset{+}{C}=N—H$	1680 (m)
Guanidinium	
monosubstituted	1630 and 1660
disubstituted	1595 and 1680
trisubstituted	1635
C—N Stretching	
Alkyl amines	1030–1230 (m)
Aryl alkyl amines	1180–1280 (m)
	1250–1360 (s)

4. FAR INFRARED

There are few data of the sort desired for characterization. The torsional methyl frequency is 269 cm^{-1} in trimethylamine and 257 cm^{-1} in dimethylamine. The trimethylamine spectrum down to 30 cm^{-1} has been studied.[8–10]

[8] W. G. Fateley and F. A. Miller, *Spectrochim. Acta* **18**, 977 (1962).
[9] V. Lorenzelli, K. D. Möller, and A. Hadni, *Compt. rend.* **249**, 239 (1959).
[10] D. W. Robinson and D. A. McQuarrie, *J. Chem. Phys.* **32**, 556 (1960).

B. Amides and Imides

1. FAR ULTRAVIOLET

In simple amides there is a band near 178 mμ ($\epsilon \sim 8000$),[11] but this is shifted to about 191 mμ for imides like succinimide ($\epsilon \sim 14,000$). The approximate doubling of the molar absorptivity is explained by the presence of two C($=$O)N linkages in the imide. Glutarimide has the maximum at 198 mμ, the bathochromic shift attributable to noncoplanarity of the imide group as the result of a twist about the CO—N bond.

In peptides (see also Chapter VIII) the several amide linkages are associated with a maximum near 185 mμ.

2. ULTRAVIOLET–VISIBLE

Acetamide has an absorption maximum at 214 mμ that may be attributed to the $n \rightarrow \pi^*$ transition of the carbonyl. In this and other respects the spectra of amides are much the same as those of the parent acids in neutral solutions.

3. NEAR INFRARED

Some similarities of amides and amines are evident since both have N—H fundamentals and overtones near each other in this region. In primary amides the asymmetric and symmetric N—H stretchings are near 2.82 and 2.92 μ and are rather strong ($\epsilon \sim 130$, in chloroform). Secondary amides are normally expected to have only one band in this region, though there is a possibility of *cis* and *trans* forms with one band for each. Tertiary amides are free from N—H stretching bands altogether.

Both primary and secondary amides have two combination bands in the vicinity of 2 μ. In many primary amides a unique 1.96 μ ($\epsilon \sim 3$) band is probably a combination of N—H stretching with the amide II or N—H deformation at 6.4 μ in the infrared.[4] Interamide hydrogen bonding has been studied in the overtone near 1.47 μ.[12]

4. INFRARED

In addition to the N—H stretching fundamental mentioned above the infrared spectra of primary and secondary amides contain as obvious

[11] D. W. Turner, *J. Chem. Soc.* **1957**, 4555.
[12] I. M. Klotz and J. Franzen, *J. Am. Chem. Soc.* **84**, 3461 (1962).

TABLE IV-3
INFRARED BANDS OF AMIDES AND RELATED COMPOUNDS

Infrared bands	Frequency (cm^{-1})
N—H Stretching	
Primary amides with free N—H	3400 and 3500 (m)
Primary amides (hydrogen bonded)	3050–3200 (several)
Secondary amides, free NH, *trans*	3400–3460 (m)
Secondary amides, free NH, *cis*	3420–3440
Secondary amides, bonded NH, *trans*	3270–3320 (m)
Secondary amides, bonded NH, *cis*	3140–3180 (m)
Secondary amides, bonded NH, *cis* or *trans*	3070–3100 (w)
Carbonyl Stretching (Amide I)	
Primary amides, solid state	1650
Primary amides, solutions	1690 (s)
Secondary amides, solid state	1630–1680
Secondary amides, solutions	1670–1700 (s)
Acetanilides, solutions[13]	1684–1706
Tertiary amides	1630–1670 (s)
Cyclic amides, large rings	1680 (s)
Cyclic amides, γ-lactams, simple	1700 (s)
Cyclic amides, γ-lactams, fused with ring	1700–1750
Cyclic amides, β-lactams, simple	1730–1760 (s)
Cyclic amides, β-lactams, fused to thiazolidine ring	1770–1780 (s)
Ureas	1660
Cyclic imides, 6-membered rings	1700 and 1710
Cyclic imides, α,β-unsaturated	1670 and 1730
Cyclic imides, 5-membered rings	1700 and 1770
Cyclic imides, α,β-unsaturated	1710 and 1790
Cyclic ureides, 6-membered	1640
Cyclic ureides, 5-membered	1720
Urethanes	1700–1735
Amide II (mainly N—H deformation)	
Primary amides, solid state	1620–1650 (s)
solutions	1590–1620 (s)
Secondary acyclic amides, solid state	1515–1570
Amide III (mainly C—N deformation)	
Secondary amides only	1290 (m)
Others	
Amide V, bonded secondary amides	720 (m) (broad)
Amide IV, secondary amides only	c.620
Amide VI, secondary amides only	c.600
Primary amides	1400–1420 (m)

[13] H. H. Freedman, *J. Am. Chem. Soc.* **82,** 2455 (1960).

components NH deformations and carbonyl bands. In addition there are a number of absorptions that cannot be (or at least have not been) allocated precisely to a single functional group. These mixture or combination bands have been numbered "amide I," "amide II," etc., even though some of them are now of understood origin (see Table IV-3). Many amide bands are quite susceptible to changes of state and substantial differences between solid and solution spectra are common.

C. Double Bond from Nitrogen to Carbon, Nitrogen, or Oxygen

The principal classes of compounds under this heading are imines and anils (C=N), azo (N=N), nitroso (N=O), and nitro (NO₂) groups, as well as the less important nitrites, nitrates, azoxy compounds, and amine oxides. Oximes belong properly to this group also but were considered as carbonyl derivatives in Chapter III.

In the ultraviolet-visible region imines and azo compounds have $n \rightarrow \pi^*$ transitions for the nitrogen electrons, and nitroso and nitro groups may additionally have similar transitions for oxygen electrons. All of these are comparable to the $n \rightarrow \pi^*$ transition of the carbonyl chromophore in being weak, particularly in relation to $\pi \rightarrow \pi^*$ bands when the group is conjugated with aromatic rings or double bonds (see Table IV-4 for nitrogen electron $n \rightarrow \pi^*$ bands).

TABLE IV-4

APPROXIMATE WAVELENGTH OF $n_N \rightarrow \pi^*$ TRANSITION
OF SOME NITROGEN FUNCTIONS[a]

Functional group	Maximum (aliphatic) (mμ)	Maximum (aromatic) (mμ)
>C=N—	210 ?	290 ?
—NO₂	270	330
—ONO₂	270	
>N—N=O	340	
—N=N—	370	400
—O—N=O	370	
—N=O	680	770

[a] All data are from reference 14 and apply to the nitrogen electrons only.

It is obvious that the infrared spectra of all these functions should have characteristic absorption in the general double bond region around 1400–1700 cm⁻¹ (see Fig. IV-3).

[14] J. W. Sidman, *Chem. Revs.* **58,** 689 (1958).

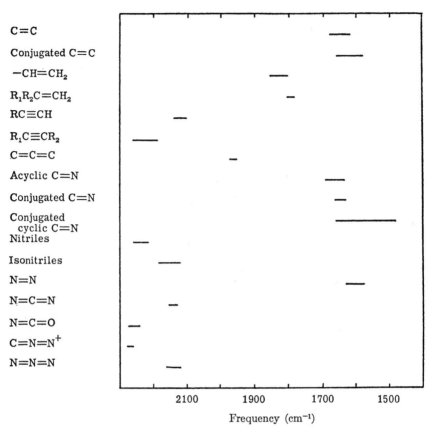

C=C

Conjugated C=C

—CH=CH₂

R₁R₂C=CH₂

RC≡CH

R₁C≡CR₂

C=C=C

Acyclic C=N

Conjugated C=N

Conjugated
    cyclic C=N
Nitriles

Isonitriles

N=N

N=C=N

N=C=O

C=N=N⁺

N=N=N

2100   1900   1700   1500

Frequency (cm⁻¹)

FIG. IV-3. Comparative double and triple bond frequencies of carbon and nitrogen (adapted from A. D. Cross, "Introduction to Practical Infrared Spectroscopy," Butterworths, London, 1960).

1. IMINES AND ANILS (C=N COMPOUNDS)

The $n \rightarrow \pi^*$ band is in the ultraviolet [for $Ph_2C=NH$, 340 mμ ($\epsilon \sim 125$)] and at shorter wavelengths than in N=N or N=O groups. The much stronger $\pi \rightarrow \pi^*$ band associated with the double bond lies in the far ultraviolet, near 190 mμ in aliphatic compounds;[15] conjugation with an aromatic ring as in the Schiff's bases or anils gives the same large bathochromic shift as in the corresponding carbon linkages and the band is in the ultraviolet. The spectrum of benzalaniline, for example,

[15] A. E. Gillam and E. S. Stern, "Electronic Absorption Spectroscopy," 2nd ed. Edward Arnold, London, 1957.

has been pronounced closely analogous to that of stilbene,[3] though the principal maximum is at shorter wavelength and the comparison is not obvious at a glance. Conjugated polyenes in which an azomethine bond (C=N) replaces one C=C, or an azine group, N—N, replaces a C—C still show the characteristic polyene spectra; in general the electronic spectra of compounds with carbon-nitrogen double bonds are very similar to the results for the corresponding carbon-carbon double bond systems.[15a]

In the infrared the identification of the C=N stretching band is often uncertain, but it usually falls at 1640–1690 cm^{-1} for open chain compounds, and at slightly lower frequencies in unsaturated conjugated systems and ring compounds. In a series of azomethines, RCH=NR', aliphatic R and R' gave the absorption band at 1664–1672 cm^{-1}; R = Ph absorbed at 1645–1650, and R = benzal at 1637–1639 cm^{-1}.[16]

2. Azo Compounds

A nitrogen-nitrogen double bond has possible $\pi \rightarrow \pi^*$ bands similar to those of carbon-carbon double bonds and also two $n \rightarrow \pi^*$ transitions, the latter generally superposed in a single absorption peak. Formation of a conjugate acid of an azo group eliminates one pair of non-bonding nitrogen electrons as does conversion of the azo to an azoxy group. Nevertheless, there are similarities in the ultraviolet spectra of stilbene, benzalaniline, azobenzene, its conjugate acid, and azoxybenzene and its conjugate acid. This has been discussed in some detail elsewhere.[3]

In simple azo compounds the $n \rightarrow \pi^*$ band is intermediate in wavelength between bands of the C=N and N=O linkages. In aromatics this absorption is stronger than in aliphatics: e.g., *cis*-azobenzene (λ_{max} = 433 mμ, $\epsilon \sim 1518$) as compared to azomethane (λ_{max} = 340 mμ, $\epsilon \sim 4.5$). The great strength of the azobenzene band is attributed to mixing of the $n \rightarrow \pi^*$ and $\pi \rightarrow \pi^*$ transitions by the distortion of the *cis*-azobenzene rings about the carbon-nitrogen bonds; since *trans*-azobenzene by virtue of its different geometry is less susceptible to such mixing, its molar absorptivity is lower.

The visible region maxima in the well known azo dyes are, of course, predominantly $\pi \rightarrow \pi^*$ transitions in extensively conjugated systems.

The infrared stretching frequency for N=N in azo groups falls in the 1575–1630 cm^{-1} region in the midst of aromatic bands and other

[15a] L. N. Ferguson, "Electron Structures of Organic Molecules," Chapter 9. Prentice-Hall, New York, 1952.

[16] F. H. Suydam, *Anal. Chem.* **35**, 193 (1963).

double bond frequencies, and is consequently not very valuable for characterization.

3. NITROSO AND OTHER NITROGEN–OXYGEN DOUBLE BOND COMPOUNDS WITH ONE OXYGEN

In the nitroso group there are $n \to \pi^*$ transitions for both the nitrogen and the oxygen, the former generally in the visible near 680 mμ ($\epsilon \sim$ 20–50)[17] and the latter, usually more intense, in the ultraviolet near 300 mμ. In N-nitrosodimethylamine the oxygen maximum undergoes a very large hypsochromic shift with increasing solvent polarity.[18] Aromatic nitroso compounds have the nitrogen transition at somewhat longer wavelengths than aliphatic, often near 750 mμ.[19]

Nitrites are generally similar to nitroso compounds in their electronic spectra, though the simple alkyl nitrites show considerable vibrational structure [20] in the 300–400 mμ region ($\epsilon \sim$ 100) and also have a band at 211–224 mμ.

For various classes of compounds with N=O groups the order of decreasing wavelength of the nitrogen $n \to \pi^*$ transition has been given as follows (see also Tables IV-4 and IV-5):

$$R_3C—NO > RS—NO > Cl—NO > R_2N—NO > RO—NO > F—NO$$

The characteristic infrared band for nitroso compounds is the N=O stretching frequency in the 1500–1600 cm^{-1} region. For tertiary nitroso groups, R_3C—NO it lies near 1550, for aromatic compounds about 1500, and for α-halo nitroso compounds as high as 1620 cm^{-1}. In nitrosamines the usual range is somewhat lower, 1430–1500 cm^{-1}, reflecting an increased contribution to the bond from a N^+=N—O^- structure.

Nitrites are characterized in the infrared by two bands in the 1610–1680 cm^{-1} region, the higher assigned to N=O stretching in the *trans* form and the lower to the *cis* form and both very strong. There are also N—O stretching bands at 750–815 cm^{-1}, and O—N=O deformation bands at 620–690 for the *trans* and 565–626 cm^{-1} for the *cis* structures.

Amine oxides have the N $\to$ O stretching band at 950–970 cm^{-1} for tertiary aliphatic amine derivatives, but at higher frequencies, 1200–1300

[17] C. Djerassi, E. Lund, F. Bunnenberg, and B. Sjöberg, *J. Am. Chem. Soc.* **83**, 2307 (1962).

[18] R. N. Haszeldine, *J. Chem. Soc.* **1954**, 691.

[19] W. J. Mijs, *Rec. trav. chim.* **77**, 746 (1958).

[20] A. Altshuller, I. Cohen, and C. M. Schwab, *J. Phys. Chem.* **62**, 621 (1958).

TABLE IV-5

Ultraviolet–Visible Spectra of Some Unsaturated Nitrogen Compounds[a]

Unsaturated nitrogen compounds	Maxima, mμ (ϵ) (solvent)
N=N	
MeN=NMe	340(4.5)(EtOH)
PhN=NMe	259.5(7800), 403.5(87)(cyclohexane)
cis-PhN=NPh	280.6(5260), 432.7(1518)(EtOH)
trans-PhN=NPh	319.6(21,300), 443.2(510) (EtOH)
N=O	
CF$_3$NO	269.5(1.8), 692.5(24)(Vapor)
PhNO	280(9000), 754.7(53)
Me$_2$NNO	232(5900), 361(125)
BuONO	222(1700), 356f(87)
BuONO$_2$	270(17)(EtOH)
MeN($\rightarrow$O)=NMe	217(7250), 274(43)(EtOH)
C=N and C≡N	
Ph$_2$C=NH	340(125)(CHCl$_3$)
MeCN	167(weak)(vapor)
NCC≡CCN	268(17.1), 280(8.4)
PhCN	222f(4.0), 270(2.7), 280(2.7)(cyclohexane)[b]
Cumulated double bonds	
EtOOCCHN$_2$	249(10,050), 377.5(16)
EtN=C=NEt	230(200), 270(25) (hexane)
PhNCO	226(4.04); 256,263,270,277(all log $\epsilon \sim$ 2.7)[b]
PhNCS	221(4.52), 270(4.03), 280(4.04)[b]

[a] Most of these data are from reference 1 and some are quite old.

[b] Log ϵ given; data from "Organic Electronic Spectral Data" Vol. IV. Interscience, New York, 1963.

cm^{-1}, in heterocyclic amines. In azoxy compounds the stretching frequency is in the 1250–1310 cm^{-1} region.

4. Nitro Groups and Nitrates

The nitro group is obviously a rather complex one in terms of number and variety of electronic transitions. The nitroparaffins have a strong pi electron transition near 210 mμ and a weak nonbonding electron excitation responsible for a maximum near 270–280 mμ ($\epsilon \sim 20$); the latter undergoes the usual hypsochromic shift in polar solvents. In nitroolefins the high intensity of the $\pi \rightarrow \pi^*$ bands in the 220–250 mμ

region appears to obscure the $n \rightarrow \pi^*$ bands entirely.[21] Nitroparaffins with two hydroxyls or one secondary amino group also present may lose through hydrogen bonding much or all of the normal 270–280 mμ maximum (see Table IV-6).[22]

TABLE IV-6

ULTRAVIOLET–VISIBLE ABSORPTION MAXIMA OF SOME NITRO COMPOUNDS[a]

Compound	Maxima, mμ (log ϵ) (solvent)
Nitromethane	260(1.58) (EtOH)
Nitroethane	260(1.63) (EtOH)
Nitropropane	270(1.59) (EtOH)
Nitrobutane	272(1.40) (EtOH)
2-Nitro-1-butanol	279(1.47) (EtOH)
$(HOCH_2)_3CNO_2$	282s(1.6) (H_2O)
$(HOCH_2)_2C(NO_2)Et$	280s(1.6) (EtOH)
2-Nitro-1-butene	223(3.01) (EtOH)
Nitrobenzene	257(3.91) (EtOH)
o-methyl derivative	250(3.77)(isooctane)
p-methyl derivative	264(4.01)(isooctane)
o-tert-butyl derivative	265s(isooctane)
p-tert-butyl derivative	265(4.03)(isooctane)
o-chloro derivative	257.5(3.89) (MeOH)
p-chloro derivative	270.5(4.03) (MeOH)
m-hydroxy derivative	226(4.04)(cyclohexane)
	262(3.76)
	319(3.34)
p-hydroxy derivative	295(4.05)(cyclohexane)
o-hydroxy derivative	230s(3.6)(cyclohexane)
	269(3.87)
	342(3.58)
p-carboxylic acid derivative	271(4.00)(H_2O)

[a] Data from Vol. IV of "Organic Electronic Spectral Data." Interscience, New York, 1963.

Nitro substitution in benzene exerts an extremely large bathochromic effect on the usual benzene spectrum, indeed, the largest effect of any simple substituent. Nitrobenzene has the primary benzene band near 260 mμ, about 60 mμ higher than in benzene, and the secondary band is completely overlapped. Since the maximum has a large bathochromic shift in polar solvents, it has been suggested that the band is perhaps produced by an intramolecular charge transfer effect rather than a shift

[21] M. J. Kamlet, J. C. Dacons, and J. C. Hoffsommer, *J. Org. Chem.* **26**, 4881 (1961).

[22] T. Urbanski, *Tetrahedron* **6**, 1 (1959).

of the benzene spectrum. Steric effects in aromatic nitro compounds have quite a complex influence on spectra;[23] both steric inhibition and enhancement of resonance have been found necessary to explain the maxima in alkylated trinitrobenzenes.

Nitrates generally differ little from nitro compounds in their electronic spectra, though the 270 mμ band of the nitroparaffins is usually present only as a shoulder in the corresponding nitrate.

Neither nitro or nitrate functions would be expected to offer any unique near infrared absorption; nevertheless, tetranitromethane has a 2.086 μ band[4] for some unknown reason.

In the infrared primary and secondary nitro groups are characterized by asymmetric and symmetric NO_2 stretching modes at 1545–1565 and 1360–1383 cm^{-1} respectively, but in tertiary nitro groups the frequencies are lower, 1530–1545 and 1345–1358 cm^{-1}. Aromatic compounds also have lower values for these bands; nitrobenzene, for example, has very strong maxima at 1523 and 1344 cm^{-1}. Monoalkylnitroethylenes have the bands at about 1524 and 1353 cm^{-1}, but di- or trialkylnitroethylenes absorb near 1515 and 1346 cm^{-1}.

The C—N stretching mode in nitro compounds is near 870 cm^{-1} and there is a C—N—O bending vibration about 610 cm^{-1}. Additional strong absorption in aromatic nitro compounds occurs near 850 cm^{-1} and sometimes also near 750 cm^{-1}.

Nitrates show the NO_2 stretching bands at 1620–1640 (asymmetric) and 1270–1285 cm^{-1} (symmetric), the C—N stretching at 855–870, an out-of-plane bending at 755–760 and NO_2 bending in the 695–710 cm^{-1} region.

D. Nitriles and Related Functions

The spectra of cumulated double bond compounds of nitrogen (iso-cyanates, isothiocyanates, azides, diazonium salts, etc.) have many similarities to those of carbon-nitrogen triple bond substances such as the nitriles and isonitriles, and are therefore briefly considered here (see Table IV-7).

1. ULTRAVIOLET–VISIBLE

By analogy with the carbon-carbon triple bond nitriles are expected to be transparent down to the far ultraviolet (e.g., acetonitrile has a maximum at 167 mμ). Conjugation, as in acetylene dinitrile, yields bands

[23] M. J. Kamlet, J. C. Hoffsommer, and H. G. Adolph, *J. Am. Chem. Soc.* **84**, 3925 (1962).

TABLE IV-7

INFRARED STRETCHING FREQUENCIES FOR CUMULENES OF NITROGEN[a]

Functional group	Frequency (cm^{-1})
Diazonium salts	2260 ± 20
Diazoketones	2060–2100
Isocyanates	2250–2275 (vs)
Carbodiimides	2130–2140 (vs)
Azides	2120–2160, **1180–1340**
Isothiocyanates	
aliphatic	1990–2140
aromatic	2040–2130
Ketenimine	c. 2000

[a] All data are from reference 7.

in the ultraviolet,[24] but there are no extended conjugated nitrile systems to compare with the poly-ynes. A cyano group in a benzene ring yields a benzenoid spectrum.

The spectra of polycyano compounds have recently become available through tetracyanoethylene and its derivatives.[25] Tetracyanoethylene itself is noted for the wide range of charge transfer bands in its spectrum as a function of solvent; a maximum at any wavelength in the visible can be arranged by proper solvent choice (see Table I-9).

The yellow color of diazomethane solutions is due to a weak ultraviolet maximum (348 mμ, $\epsilon \sim 4.5$) and there is also a shorter wavelength band of higher intensity.[15] Aliphatic azides generally have two maxima in the ultraviolet (e.g., EtN$_3$, 222 and 287 mμ, both weak).

2. NEAR INFRARED

A considerable number of isocyanates have the first overtone of the C=O vibration at 2.642 μ ($\epsilon \sim 1$–3), and this, in conjunction with detailed analysis of C—H and combination bands in this region of the spectrum, has been proposed as a useful correlation.[26]

3. INFRARED

Nitriles are best characterized by the strong triple bond stretching vibration near 2240–2260 for saturated aliphatic nitriles, 2220–2240 for

[24] F. Miller and R. Hannan, *Spectrochim. Acta* **12**, 321 (1958).
[25] B. C. McKusick, R. E. Heckert, T. L. Cairns, D. D. Coffman, and H. F. Mower, *J. Am. Chem. Soc.* **80**, 2806 (1958).
[26] D. J. David, *Anal. Chem.* **35**, 37 (1963).

aromatic, and 2215–2235 cm^{-1} for α,β-unsaturated compounds. The intensity varies with substituents, the more polar groups yielding the highest intensities and aromatics absorbing more strongly by far than aliphatics.

Isonitriles in chloroform have the stretching frequency near 2140 cm^{-1}. In metal complexes of these a displacement of 60–200 cm^{-1} toward lower frequencies may occur.[27]

The nitrogen cumulenes have strong bands in the 2000–2300 cm^{-1} region (listed in Table IV-7) and weak bands near 1350 cm^{-1}. These are respectively the asymmetric and symmetric stretchings of the X=Y=Z group, where X, Y, and Z are carbon, nitrogen, and oxygen. Except for atmospheric carbon dioxide at 2349 cm^{-1} there are very few functions other than nitrogen cumulenes and triple bonds absorbing appreciably in the 2000–2300 cm^{-1} vicinity.

Isocyanates, RNCO, have the asymmetric stretching band at 2250–2275 cm^{-1} and isothiocyanates a strong broad doublet in the 2100 cm^{-1} region with a molar absorptivity near 700.[28]

Table IV-8 summarizes the most useful regions of the spectrum for many of the nitrogen functions discussed in this chapter.

TABLE IV-8

Most Distinctive Regions of the Spectrum for Nitrogen Functions[a]

Function	Best region(s)	Distinctive features
Aliphatic amines	Infrared	Effect of hydrochloride formation
Primary and secondary	Near infrared	N—H stretching
Aralkyl and aryl amines	Ultraviolet	Changes in acid, steric effects
Amides and peptides	Infrared, far ultraviolet	Amide I-VI bands
Azomethines	Ultraviolet	
Imines	Near infrared	
Azo groups	Ultraviolet–visible	*Cis-trans* differences, conjugate acid spectra
Nitroso groups	Visible	Weak, long wavelength band
Nitro groups	Ultraviolet	Steric effects (aromatic)
Amine oxides	Infrared	
Nitriles and isonitriles	Infrared	C≡N region
Isocyanates, isothiocyanates	Infrared	Cumulated double bond stretching
Azides	Infrared	

[a] These statements are opinions.

[27] F. A. Cotton and F. Zingales, *J. Am. Chem. Soc.* **83**, 351 (1961).

[28] N. S. Ham and J. B. Willis, *Spectrochim. Acta* **16**, 279 (1960).

V

Heterocycles

Only the simpler heterocyclic compounds containing oxygen, nitrogen, or sulfur as the hetero atom can be considered here. From the enormous size of the *Ring Index* the difficulties involved in a brief account of heterocyclic compound spectra may be readily deduced, and the recent publication of a 350-page volume (edited by A. R. Katritzky) on the spectroscopy of heterocyclic compounds certainly emphasizes the point.

Saturated heterocycles, or even those containing one double bond in a 5- or 6-membered ring, offer little of spectrophotometric interest that cannot be predicted from the acyclic analogs of these substances. Oxygen-containing saturated rings are like ethers, and in their electronic spectra should have bands below 200 mμ only. Nitrogen-containing saturated rings resemble acyclic amines and their electronic spectra show absorption bands at the low end of the ordinary ultraviolet wavelength range that are lost in acid solutions. Absorption bands for a few saturated heterocycles are included in Table V-1. Only in the far infrared where over-all ring frequencies occur can these compounds have unique spectra.

The most important heterocycles are those having aromatic character. The simplest 5-membered rings of this type are the furans, pyrroles, and thiophenes (see Table V-1 for ultraviolet spectra). However, the most intensive spectroscopic studies have been performed in the 6-membered nitrogen ring compounds such as the pyridines, pyrimidines, and condensed ring derivatives of these.

A. Aza Benzenes and Naphthalenes (Pyridines, Pyrimidines, Quinolines, etc.)

1. ULTRAVIOLET–VISIBLE

The replacement of one or more carbons in the benzene or naphthalene rings by nitrogen leaves the three major absorptions (α, p, and β bands, in order of decreasing wavelength) still recognizable in neutral solvents. As with benzene the shortest wavelength band may be in the far ultra-

TABLE V-1
FAR ULTRAVIOLET AND ULTRAVIOLET MAXIMA
OF SOME SINGLE RING HETEROCYCLES[a]

Single ring heterocycles	Maxima, mμ (log ϵ)(solvent)
Saturated rings	
Ethylene oxide	143.5, 157.2, 171
Tetrahydrofuran	170, 190
Tetrahydropyran	175, 189
Dioxan	188
Pyrrolidine	171 (3.4), 196 (3.3)
Piperidine	198f (3.5)
Piperazine	196 (3.7)
5-Membered aromatic rings (two double bonds)	
Furan	191, 205
Pyrrole	172, 183, 211 (4.5), 340 (2.5)
Thiophene	215 (3.8), 231 (3.8) (EtOH)
Selenophene	232 (3.56), 249 (3.75) (EtOH)
Imidazole	207 (3.70) (EtOH)
Pyrazole	210 (3.50) (EtOH)
Isoxazole	211 (3.60) (EtOH)
Thiazole	240 (3.60) (EtOH)
1,2,3-Triazole	210 (3.60) (EtOH)
6-Membered aromatic rings[b]	
Pyridine	251 (3.30), 270s (2.65)
Pyridazine	246 (3.11), 340 (2.50)(cyclohexane)
Pyrimidine	243 (3.31), 298 (2.51)(cyclohexane)
Pyrazine	260 (3.74), 328 (3.02)(cyclohexane)
sym-Triazine	222 (2.18), 272 (2.95)(cyclohexane)
sym-Tetrazine	252 (3.33), 542 (2.92)(cyclohexane)
γ-Pyran[c]	222 (3.85), 238 (3.71) (MeOH)

[a] These data were compiled from references 1–3; see also Tables V-2 and V-5.

[b] Only the two longest wavelength bands listed.

[c] From J. Strating, J. H. Keijer, E. Molenaar, and L. Brandsma, *Angew. Chem.* **74,** 465 (1962).

[1] H. H. Jaffé and M. Orchin, "Theory and Applications of Ultraviolet Spectroscopy," Chapter 14. Wiley, New York, 1962.

[2] C. N. R. Rao, "Ultra-Violet and Visible Spectroscopy," Chapter 6. Butterworths, London, 1961.

[3] S. F. Mason, *Quart. Revs. (London)* **15,** 330 (1961).

violet; e.g., pyridine at 178 mμ.[4] The decreased symmetry introduced by the nitrogen will increase the intensity and reduce the fine structure of the α band (near 255 mμ in benzene), and an additional longer wavelength absorption identified with the $n \rightarrow \pi^*$ transitions of the nitrogen electrons may be detected (see Table V-1), though only as a shoulder in some cases.

The effect of substituents on the α band near 250 mμ in pyridine is complicated by the basicity of the nitrogen as well as the existence of three unequivalent positions for substitution. As in benzene most substituents exert a bathochromic effect, but a 3-substituent has a greater effect than a 2-substituent and 4-substitution is the least effective. In acidic solvents the formation of a cationic species uses the free electron pair on the ring nitrogen, and the $n \rightarrow \pi^*$ transition at longer wavelength is lost while the α band increases in intensity. If the substituent is OH, SH, COOH, or another acidic function, the molecule may exist in a neutral form, a cation, an anion, or a zwitter ion, depending largely on the acidity of the solution; the wavelength of the α band for these four forms increases in the order as given from neutral molecule to zwitter ion.[3] Sulfur-containing substituents have a larger bathochromic effect than either oxygen or nitrogen.

N-Alkylpyridinium compounds should have spectra similar to those of the parent pyridines in acidic solvents but with some bathochromic shift. In addition, however, there may be strong charge-transfer bands in N-alkylpyridinium halides between the halogen and the ring,[5] and in appropriately substituted derivatives intramolecular charge transfer bands as well. These charge-transfer bands are in the ultraviolet, varying greatly in position with the polarity of the solvent used.

Mason[3] has found very great similarities in the ultraviolet spectra of pyrimidines, pyridazines, pyrazines, and higher poly-aza aromatics as compared to pyridine. There are differences in basicity and in number of nitrogen lone-pair electrons that have some effect, and hydroxyl or thiol groups substituted in appropriate positions may cause a loss of the dominant aromatic character of the ring through formation of lactam and related structures; for such compounds the spectra may be far from those of typical pyridines.

The nitrogen analogs of naphthalene such as quinoline, isoquinoline, cinnoline, etc., also have spectra resembling that of naphthalene, at least in neutral solvents. In higher condensed ring systems the analogy

[4] D. W. Turner *in* "Determination of Organic Structures by Physical Methods," Chapter 5. (F. C. Nachod and W. D. Phillips, eds.), Academic Press, New York, 1962.
[5] E. M. Kosower and J. A. Skorcy, *J. Am. Chem. Soc.* **82**, 2195 (1960).

TABLE V-2

ULTRAVIOLET MAXIMA OF SOME SIMPLE SUBSTITUTED AZA AROMATICS[a]

Substituted aza aromatics	Maxima, mμ (ϵ)	Solvent
Substituents in pyridine		
2-Et	262 (3690)	H_2O
3-Et	263 (3210)	H_2O
4-Et	255 (2150)	H_2O
2-OMe	269 (3230)	H_2O
3-OMe	276 (2960)	H_2O
4-OMe	235 (2000)	H_2O
2-Cl	265 (2920)	Isooctane
3-Cl	268 (2400)	Isooctane
Substituents in pyridazine[b]		
3-Me	251(3.11), 310(2.60)	EtOH
4-Me	245(3.12), 250(3.14), 303(2.54)	EtOH
3,4-Me$_2$	253(3.23), 302(2.54)	EtOH
Substituents in pyrimidine		
2-COOH	246(2120), 277(374)	EtOH
4-COOH	256(3820), 303(295)	EtOH
5-COOH	247(1870), 280(561)	EtOH
Substituents in quinoline		
2-Me	279(3200), 315(3860)	10% EtOH
3-Me	286(3350), 318(3120)	10% EtOH
4-Me	283(4530), 313(2730)	10% EtOH
6-Me	285(2340), 317(2330)	10% EtOH
7-Me	292(2170), 318(2400)	10% EtOH
8-Me	292(3750), 314(2650)	10% EtOH
2-Cl	282(3320), 318(4610)	10% EtOH
3-Cl	283(3060), 323(3560)	10% EtOH
5-Cl	292(4640), 317(3170)	EtOH
2-OH[b]	230(4.55), 268(3.82)	95% MeOH
3-OH[b]	286(3.33), 322(3.59), 333(3.64)	95% MeOH
8-OH[b]	243(4.54), 310(3.40)	95% MeOH

[a] All maxima are not necessarily listed.

[b] Log ϵ is given in place of ϵ.

of the ultraviolet spectrum of the aza derivative to the parent hydrocarbon is generally preserved as well; e.g., benzo[c]cinnolines have spectra like phenanthrenes.[6]

[6] J. F. Corbett, P. F. Holt, A. N. Hughes, and M. Vickery, *J. Chem. Soc.* **1962**, 1812.

The N-oxides of the aza aromatics in nonpolar solvents undergo a considerable bathochromic and hyperchromic shift, possibly because of the lengthening of the conjugated system. In hydroxylic solvents where hydrogen bonding of the oxygen may occur, this bathochromic shift relative to the parent aza aromatic virtually disappears, though the hyperchromic effect persists.

For characterization purposes it is clearly desirable to determine the spectra of the aza aromatics at several pH values. Simple bases like pyridine and quinoline will have different spectra in acid and neutral solutions, and the possible presence of acidic substituents in the ring will allow further changes in going from neutral to basic solutions. The poly aza aromatics may undergo multiple changes of spectrum with pH, and some of the common hydroxypyrimidines, for example, may show a noticeable change in spectrum for every 1–2 unit change in pH (see Section E of Chapter VIII).

In Table V-2 are listed the principal maxima in the ultraviolet of a few substituted aza aromatics. Lists of the longest wavelength band for over 200 polynuclear heterocyclic aromatics have been compiled.[7]

2. Infrared

Considerable data for pyridines, pyrimidines and quinolines have been compiled, but many aza aromatics remain little studied. In most respects the spectra are similar to those of the corresponding aromatic hydrocarbons except where substituents have destroyed part or all of the aromatic characteristics of the molecule (e.g., many hydroxy and amino pyrimidines). The principal differences are said to be in the hydrogen deformation region.[8]

The C—H stretching bands in the 3000–3100 cm^{-1} region generally are like those of benzene. In pyridinium ions a $-\overset{+}{N}-$H frequency is also present, often accounting for one or more bands in the 1900–2200 cm^{-1} range.

In the 1650–2000 cm^{-1} region there are overtones and combinations of out-of-plane C—H bending modes that depend on the type of substitution for the pattern of bands produced.

The most detailed analysis has been applied[9] to the 1350–1600 cm^{-1}

[7] C. Karr, Jr., *Appl. Spectry.* **14**, 146 (1960).

[8] L. J. Bellamy, "The Infra-red Spectra of Complex Molecules, 2nd ed., Chapter 16. Wiley, New York, 1958.

[9] A. R. Katritzky, *Quart. Revs.* (*London*) **13**, 353 (1959).

TABLE V-3

CHARACTERISTIC BANDS IN THE INFRARED SPECTRA (BELOW 1650 CM⁻¹)
OF 6-MEMBERED RING NITROGEN HETEROCYCLES[a]

Characteristic bands	(cm⁻¹) Frequencies
1350–1600 cm⁻¹ Bands for	
Pyridine	1599, 1583, 1482, 1441
2-Substituted pyridines	1615→1585,[b] 1572, 1471, 1433
3-Substituted pyridines	1595, 1577, 1465←1485, 1421
4-Substituted pyridines	1603, 1561, 1480←1520, 1415
Polysubstituted pyridines	1597–1610, 1564–1588, 1490–1555
Pyridine-1-oxide	1612, 1468
3-Substituted pyridine-1-oxides	1605, 1563, 1480, 1434
Pyridazine	1572, 1565, 1444, 1414
Pyrimidine	1610, 1569, 1461, 1400
Substituted pyrimidines	1555–1590, 1520–1565, 1400–1480, 1375–1410
Pyrazine	1584, 1523, 1490, 1418
1,3,5-Triazine	1556, 1410
Ring-breathing modes near 1000 cm⁻¹ for	
Pyridine	1030, 993
2-Substituted pyridines	994
3-Substituted pyridines	1025
4-Substituted pyridines	993
Pyrimidines	990
Pyrazine	1022
Out-of-plane modes for	
Pyridines with	
5 adjacent hydrogens	749, 700
4 adjacent hydrogens	740–780
1,2,3,5-hydrogens	770–820, 690–730
1,2,4,5-hydrogens	790–850
Azines with[c]	
4 adjacent hydrogens	760
1,2,3,5-hydrogens	721, 680
1,2,4,5-hydrogens	804

[a] See also Table V-4 following; all data are from reference 9.

[b] The arrow indicates that electron-donor substituents are at the upper end and acceptors at the lower end of the range indicated.

[c] Nitrogen heterocycles with more than one ring nitrogen.

region where the aromatic ring stretching frequencies are located. There are usually four bands, near 1605, 1575, 1480, and 1430 cm^{-1} (see Table V-3). For monosubstituted compounds the 1605 cm^{-1} intensity is high for electron-donating groups, low for weakly interacting substituents and high for electron accepting groups; this is true of monosubstituted benzenes, 3-substituted pyridines and 4-substituted pyridine-1-oxides. In 2- and 4-substituted pyridines electron acceptors give low intensity instead. The band near 1575 cm^{-1} is usually weaker than the 1605 cm^{-1} band but varies similarly with substituents. The 1480 cm^{-1} absorption is strong for electron-donor substituents and weak or missing if these are absent. The 1430 cm^{-1} band is relatively independent of the nature of any substituent.

There are also ring-breathing modes near 1000 cm^{-1} and out-of-plane C—H deformations in the 700–900 cm^{-1} region (Table V-3).

The in-plane C—H deformations near 1200 cm^{-1} for pyridines vary with substitution pattern much as in substituted benzenes (Table V-4),

TABLE V-4

IN-PLANE HYDROGEN DEFORMATION MODES FOR HETEROAROMATICS[a]

Compound	Frequency (cm^{-1})	Benzene analog
Pyridine	1218, 1148, 1085, 1068	Monosubstituted
2-Substituted pyridines	1279, 1147, 1093, 1048	*o*-Disubstituted
3-Substituted pyridines	1190(?), 1124, 1103, 1038	*m*-Disubstituted
4-Substituted pyridines	1220, 1067	*p*-Disubstituted
Pyridazine	1160, 1063	*o*-Disubstituted
Pyrimidine	1220, 1165, 1140, 1021	*m*-Disubstituted
Pyrazine	1148, 1067	*p*-Disubstituted
Furan	1270, 1067	*o*-Disubstituted
Pyrrole	1076, 1046, 1015	*o*-Disubstituted
Thiophene	1283, 1077, 1032, 909(?)	*o*-Disubstituted
2-Substituted furans	1220, 1158, 1076	vicinal-trisubstituted
2-Substituted thiophenes	1081, 1043	vicinal-trisubstituted

[a] All data are from reference 9; see also Table V-3.

except that a monosubstituted pyridine resembles a disubstituted benzene, a disubstituted pyridine is comparable to a trisubstituted benzene, etc.

Substituents that have characteristic infrared absorptions of their own are not greatly influenced by the attachment of the heterocyclic nucleus, unless there is interaction with the hetero atom. A 2-hydroxy-pyridine, for example, is largely in the α-pyridone form, with a concomitant carbonyl stretching band in the 1650–1690 cm^{-1} region and a N—H band at 3400 cm^{-1}.

A large number of articles dealing with the infrared spectra of many derivatives of a single heterocycle have been published, including the following representative examples pertinent to this section: *sym*-triazines,[10,11] pyridones and thiopyridones,[12] hydroxy pyridines and pyrimidines,[13] monosubstituted quinolines[14] and acridines.[15] See also Katritzky.[16]

B. Conjugated 5-Membered Heterocyclic Rings

Furans, pyrroles, and thiophenes containing, respectively, oxygen, nitrogen, and sulfur as the single hetero atom are the simplest examples, but two hetero atoms per ring (e.g., imidazoles, pyrazoles, isoxazoles, thiazoles, etc.) as well as fusion of the 5-membered ring to benzene (e.g., benzofurans, indoles, etc.) create a multitude of additional heterocycles that are considered primarily aromatic.

1. ULTRAVIOLET–VISIBLE

The spectra of furans, pyrroles, and thiophenes are considered roughly analogous to those of benzene derivatives. There are fairly intense bands near 200 mμ and what might be considered benzenoid bands at longer wavelengths. For this sort of comparison the hetero atom may be considered the equivalent of a —CH=CH— group, a viewpoint especially helpful in classifying the spectra of fused ring systems. For example, the spectra of benzofuran, indole, thianaphthene, and benzimidazole (all 5-membered rings fused to benzene) are like those of naphthalene, and fluorenes, and carbazoles are comparable to anthracene.

It is difficult to draw any useful general conclusions about the electronic spectra of these rings beyond those above. Substituent effects on the individual systems have been studied, and some of these data are summarized in Table V-5.

[10] H. K. Reinschuessel and N. T. McDevitt, *J. Am. Chem. Soc.* **82,** 3756 (1960).

[11] W. M. Padgett and W. F. Hamner, *J. Am. Chem. Soc.* **80,** 803 (1958).

[12] A. R. Katritzky and R. A. Jones, *J. Chem. Soc.* **1960,** 2947.

[13] A. Albert and E. Spinner, *J. Chem. Soc.* **1960,** 1221, 1226, 1232.

[14] A. R. Katritzky, *J. Chem. Soc.* **1960,** 2943.

[15] A. I. Gurevich, *Opt. Spectry. (USSR) (English Transl.)* **12,** 20 (1962).

[16] A. R. Katritzky, ed., "Physical Methods in Heterocyclic Chemistry," Vol. II, Academic Press, New York, 1963.

TABLE V-5

SUBSTITUENT EFFECTS ON ULTRAVIOLET MAXIMA
OF 5-MEMBERED HETEROCYCLES[a]

Substituents	Maxima, mμ (log ε)	Solvent
Substituents in furan		
2-CH₃	None above 220 mμ	EtOH
2-CH₂OH	None above 220 mμ	EtOH
2-CHO	227(3.48), 272(4.12)	EtOH
2-COCH₃	225(3.38), 269(4.13)	EtOH
2-COOH	214s(3.58), 243(4.03)	EtOH
3-COOH	232(3.36)	EtOH
2-CH=CH₂	260.5(3.97)	H₂O
2-NO₂	225(3.53), 315(3.91)	H₂O
2,5-(NO₂)₂	230(3.84), 310(4.06)	H₂O
Substituents in thiophene		
2-Br	237(3.96)	EtOH
3-Br	232(3.61)	EtOH
2-CH₃	234(3.58)	H₂O
3-CH₃	233(3.8)	—
2-CHO	260(4.04), 286(3.86)	EtOH
3-CHO	251(4.12)	EtOH
2-COOH	246(3.93), 264s(3.76)	EtOH
3-COOH	241(3.92)	EtOH
2-CONH₂	248(3.78), 272(3.87)	EtOH
3-CONH₂	241(3.89)	EtOH
2-NO₂	270(3.80), 296(3.78)	EtOH
2-CH=CH₂	275(3.96)	EtOH
Substituents in pyrrole		
N-COCH₃	238.5(4.03), 288s(2.88)	EtOH
2-COCH₃	250(3.64), 287(4.20)	EtOH
2-CHO	252(3.70), 299(4.22)	EtOH
2-COOH	228(3.65), 258(4.10)	EtOH
N-CH₃-2-COOH	236(3.80), 261(4.09)	EtOH
Substituents in imidazole		
N-CH₃	212(3.63)	EtOH
4-CH₃	215(3.67)	EtOH
Substituents in isoxazole		
3-CH₃	217(3.74)	EtOH
4-CH₃	222(3.50)	EtOH
5-CH₃	213(3.19)	EtOH
3,4-(CH₃)₂	219(3.52)	H₂O
4,5-(CH₃)₂	226(3.68)	H₂O

[a] All data are from "Organic Electronic Spectral Data" Vol. IV. Interscience, New York, 1963.

Changes of spectrum with pH in these compounds are relatively unimportant. Pyrrole is feebly acidic. Only the nitrogen heterocycles in this group are capable of substitution at the hetero atom.

2. INFRARED

The similarity to aromatic hydrocarbons is apparent in the infrared spectra of these compounds also.[9] The C—H stretching vibrations are near 3000–3100 cm^{-1}, and the pyrroles have N—H stretching bands at 3400–3450 cm^{-1} ($\epsilon \sim 120$) in solutions. In the characteristic ring-stretching region from 1350 to 1600 cm^{-1} there are usually only three

TABLE V-6

CHARACTERISTIC INFRARED BANDS OF 5-MEMBERED RING HETEROAROMATICS[a]

Heteroaromatics	Frequency (cm^{-1}) (approx.)
Ring-stretching region for	
2-Substituted furans	1570←1605, 1475←1510, 1380–1400[b]
Polysubstituted furans	c. 1560, c. 1510
2-Substituted thiophenes	1523, 1442, 1354, 1231
3-Substituted thiophenes	c. 1530, c. 1410, c. 1370
Substituted pyrroles	c. 1565, c. 1500
Thiazole	1615, 1485, 1385
Substituted thiazoles	c. 1610, c. 1500, c. 1380
Imidazole	1550, 1492, 1451
1,2,3-Triazole	1520, 1450, 1410
Substituted isoxazoles	c. 1600, c. 1460, c. 1420
Ring breathing region for	
Furan	994 ($\epsilon \sim 170$)
2-Substituted furans	1015 ($\epsilon \sim 85$)
Thiophene	832 ($\epsilon \sim 95$)
2-Substituted thiophenes	823 ($\epsilon \sim 50$)
Out-of-plane modes for	
Furan	837, 744
2-Substituted furans	925, 884, c. 800
Thiophene	832, 710
2-Substituted thiophenes	925, 853, 800
Pyrrole	838, 768

[a] All data from reference 9; see also Table V-4.

[b] The arrow means electron-donor substituents absorb at high end and acceptors at low end of range indicated.

bands, near 1590, 1490, and 1400 cm^{-1} (see Table V-6) and electron-withdrawing substituents increase the intensity of all three.

The in-plane hydrogen deformations (see Table V-4) in the 1000–1250 cm^{-1} range are comparable to those of disubstituted benzenes or monosubstituted pyridines for furan, pyrrole, and thiophene, and monosubstituted 5-membered ring heterocycles should be compared to trisubstituted benzenes, and so on.

Ring breathing modes near 1000 cm^{-1} and out of plane deformations for C—H linkages are also summarized in Table V-6.

Among the many papers devoted to substituent effects in heterocyclic spectra the following are pertinent examples here: 2-substituted furans,[17] 2-substituted thiophenes,[18] isoxazoles,[19] pyrazoles,[20] 5-membered rings fused to benzene,[21] and benzo-1,2,3-triazoles.[22]

[17] A. R. Katritzky and J. M. Lagowski, *J. Chem. Soc.* **1959**, 657.
[18] A. R. Katritzky and A. J. Boulton, *J. Chem. Soc.* **1959**, 3500.
[19] A. R. Katritzky and A. J. Boulton, *Spectrochim. Acta* **17**, 238 (1961).
[20] G. Zerbi and C. Alberti, *Spectrochim. Acta* **18**, 407 (1962).
[21] D. G. O'Sullivan, *J. Chem. Soc.* **1960**, 3278.
[22] D. G. O'Sullivan, *J. Chem. Soc.* **1960**, 3653.

VI

Organic Compounds Containing Halogen, Sulfur, Phosphorus, Silicon, or Boron

A. Organic Halides

Substitution of hydrogen by halogen adds nonbonding electrons, with the resultant possibility of $n \rightarrow \sigma^*$ electron transitions that usually appear as low intensity absorption bands in the ultraviolet. In the infrared stretching bands for carbon-halogen bands occur at quite low frequencies because of the relatively large atomic masses of the halogens. Both ultraviolet and infrared bands associated with halogen generally progress to longer wavelength with increasing atomic weight of the halogen, i.e., in the order, fluorine, chlorine, bromine, and iodine.

Generally fluorine alters the electronic spectrum very little as compared to the parent hydrocarbon, though polyfluoro compounds may exhibit a hypsochromic shift. The high electronegativity of fluorine accounts for some rather large changes in characteristic infrared stretching frequencies, however, as well as very high intensity in the C—F stretching bands themselves.

Iodine is heavy enough to permit possible expansion of its valence shell and other complexities in its electronic transitions. The interpretation of the ultraviolet spectra of aromatic iodides and iodonium compounds, for example, offers difficulties not expected in compounds of other halogens.

1. Far Ultraviolet

Below 200 mμ the alkyl halides are characterized by the following progression of absorption bands[1]: R—F, under 100, R—Cl, under 165; R—Br, about 175 ($\epsilon \sim 5000$); and R—I, 170 ($\epsilon \sim 10,000$) and 190 mμ ($\epsilon \sim 6000$). The weak absorption in methyl chloride at 172.5 mμ is probably the $n \rightarrow \sigma^*$ transition that falls in the normal ultraviolet for bromides and iodides.

[1] D. W. Turner *in* "Determination of Organic Structures by Physical Methods" (F. C. Nachod and W. D. Phillips, eds.), pp. 339–400. Academic Press, New York, 1962.

Aryl halides show the normal benzene band in the 185–196 mμ range (see Table II-9 for some examples). In hexachlorobenzene this band is shifted to 216 mμ, an indication that polyhalogenation generally yields a bathochromic shift. However, perfluorotoluene and other polyfluorobenzenes exhibit a small hypsochromic shift instead.

2. Ultraviolet–Visible

Methyl bromide vapor has a weak 204 mμ ($\epsilon \sim 200$) band and methyl iodide in petroleum ether absorbs at 257.5 mμ ($\epsilon \sim 265$)[2]; both bands are $n \to \sigma^*$ transitions, as indicated by a hypsochromic shift with increasing solvent polarity.[3] Higher alkyl halides have similar absorption, though chain branching is usually associated with a small shift to longer wavelengths.[3] Substitution of more than one halogen on a single carbon is associated with a bathochromic shift also, and in the polyiodides there may be several bands as the result of orbital splitting by iodine-iodine interaction (e.g., CH_2I_2, $\lambda\lambda_{max}$ 212, 240, 290 mμ; CHI_3, $\lambda\lambda_{max}$ 274, 307, 349 mμ).

In conjugated dienes substitution of chlorine or bromine for one hydrogen increases λ_{max} by approximately 17 mμ,[4] and an α-bromo substituent in α,β-unsaturated carbonyl compounds increases λ_{max} by about 23 mμ; β-bromo compounds undergo only a 10–12 mμ bathochromic shift.

Except for the iodo compounds aryl halides have spectra quite close to those of the parent hydrocarbons. In the aryl iodides the band at 226 mμ (log $\epsilon = 4.12$) in iodobenzene appears to be a specific characteristic of the iodine, since all of them absorb in this region. Excitation of the iodine lone-pair electrons may account for the band, though its intensity is extremely high for such an assignment. The diphenyliodonium salts[5] have spectra much like that of iodobenzene except that the benzenoid band near 260 mμ is missing.

3. Near Infrared

In the alkyl halides all significant absorption in this region can be assigned to C—H overtones and various combinations,[6] as there is no characteristic absorption for halogen here.

[2] S. F. Mason, *Quart. Revs. (London)* **15**, 287 (1961).

[3] H. Kimura and S. Nagakura, *Spectrochim. Acta* **17**, 166 (1961).

[4] H. H. Jaffé and M. Orchin, "Theory and Applications of Ultraviolet Spectroscopy," p. 218. Wiley, New York, 1962.

[5] F. M. Beringer and I. Lillien, *J. Am. Chem. Soc.* **82**, 5135 (1960).

[6] R. J. W. LeFèvre, R. Roper, and A. J. Williams, *Australian J. Chem.* **12**, 743 (1959).

4. Infrared

The C—F stretching frequency may vary considerably as a function of other structural features in the molecule (see Table VI-1) and in turn may affect other group frequencies, but most of the simple fluorine compounds have an extremely strong band in the 1000–1100 cm^{-1} region

TABLE VI-1
INFRARED FREQUENCIES OF FLUORINE BONDS AND BONDS AFFECTED BY FLUORINE[12]

Structure[a]	Stretching vibration of	Frequency (cm^{-1})
—CF$_2$CF$_2$—	C—F	1150, 1210, 1450
CF$_3$CF$_2$—	C—F(CF$_3$)	1050
—CF(CF$_3$)CF$_2$—	C—F(CF$_3$)	970
R$_F$CH$_2$OH	O—H	3300
R$_F$CHF$_2$	C—H	3000
R$_F$CF=CF$_2$	C=C	1800
R$_F$CF=CFR$_F$	C=C	1735
(R$_F$)$_2$C=CF$_2$	C=C	1750
(R$_F$)$_2$C=O	C=O	1785
R$_F$COCH$_2$—	C=O	1770
R$_F$COF	C=O	1880
R$_F$COCl	C=O	1810
R$_F$COOH	C=O	1775
R$_F$CH$_2$COOH	C=O	1720
(R$_F$CO)$_2$O	C=O	1820, 1890

[a] R$_F$ stands for a fully fluorinated alkyl group.

and all may be expected to have strong bands somewhere in the 1000–1400 cm^{-1} range. The high intensity is more characteristic of fluorine than the exact frequency of the bands. As the result of conformational effects a single C—F linkage may give rise to more than one stretching band, and polyfluorinated compounds have multiple bands in the 1100–1400 cm^{-1} range. The CF$_3$CF$_2$-group is fairly uniquely characterized by bands at 1325–1365 and 730–745 cm^{-1}.

For C—Cl groups the stretching frequency is in the 600–800 cm^{-1} range, generally 700–750 cm^{-1} for a single chlorine but higher for polychloro compounds. In chlorinated cyclohexanes a 742 cm^{-1} band for equatorial and a 688 cm^{-1} band for axial conformations are among the best known relationships of conformation to physical properties. In

[12] G. H. Beaven, E. A. Johnson, H. A. Willis, and R. G. J. Miller, "Molecular Spectroscopy," p. 251. Macmillan, New York, 1961.

general, an equatorial C—Cl is represented by a higher frequency than an axial.

Except for fluorine the effect of halogen substitution at a carbon-carbon double bond is a decrease in the double bond stretching frequency with increasing atomic weight of the halogen.[7] In aromatic halides the position of substitution may be deduced from the band pattern in the 1100 cm^{-1} region[8]; thus p-substituted chlorine, bromine or iodine respectively are responsible for 1089–1098[6] ($\epsilon \sim 175$), 1068–1073 ($\epsilon \sim 130$), and 1057–1061 cm^{-1} ($\epsilon \sim 50$) bands, and similar regularities for *ortho* and *meta* substitution are known.

5. FAR INFRARED

Most of the C—Br and C—I stretching frequencies are in this region because of the large mass of these halogens. The C—Br linkage is identified in solution spectra with bands around 650 and 560 cm^{-1}, and the equatorial conformation is associated with the 700–750 cm^{-1} region while the axial has a band in the 550–690 cm^{-1} range.[9]

The C—I stretching is at still lower frequency, generally the 500–600 cm^{-1} range.

The considerable theoretical interest attached to the alkyl halides has provided spectral data on these as far as 30 cm^{-1} and even in the crystalline state.[10,11]

B. Sulfur Compounds

The extremely large variety of these compounds makes it hard to do more than mention the principal types.

1. ULTRAVIOLET–VISIBLE

Expansion of the valence shell in sulfur accounts for the formation of sulfoxides, sulfones and a number of other unique types of compounds and also allows a conjugation of two chromophoric groups linked by a sulfur atom. The easier transitions of sulfur electrons as compared to oxygen should make analogous absorption bands appear at longer wavelengths in sulfur compounds. Spectra of several hundred sulfur

[7] R. P. Bauman, "Absorption Spectroscopy," p. 295. Wiley, New York, 1962.

[8] A. R. Katritzky and J. M. Lagowski, *J. Chem. Soc.* **1960**, 2421.

[9] A. D. Cross, "Introduction to Practical Infra-Red Spectroscopy." Butterworths, London, 1960.

[10] W. J. Lafferty and D. W. Robinson, *J. Chem. Phys.* **36**, 85 (1962).

[11] S. Palik and C. Rao, *J. Chem. Phys.* **26**, 1401 (1957).

compounds have been collected as an appendix of the Price and Oae book "Sulfur Bonding" (Ronald Press, New York, 1962).

a. Thiols.

The lower aliphatic members of this series have an absorption band near 227 mμ, perhaps the $n \rightarrow \sigma^*$ transition of free sulfur electrons. Thiophenol has an absorption band near 236 mμ and another with considerable vibrational structure in the 260–295 mμ region; it is possible that these are the 200 and 255 mμ band systems of benzene shifted to longer wavelengths, though it has also been suggested that the longer band could be a $n \rightarrow \pi^*$ transition instead.[13]

b. Sulfides.

Although these are the formal sulfur analogs of ethers, their spectra are vastly more complex. The simple dialkyl sulfides have a band near 210 mμ ($\epsilon \sim$ 1000–2500) and a shoulder at 229 mμ ($\epsilon \sim$ 100); the 210 mμ band shifts slightly to higher wavelengths and greater intensity with increased branching of the alkyl groups. In β,γ-unsaturated sulfides there is a shift to about 221 mμ as the result of what might be a hyperconjugation-like interaction of sulfur with the double bond. In β-keto-sulfides there appears to be a similar effect. The α,β-unsaturated sulfides are conjugated systems and have $\pi \rightarrow \pi^*$ bands (at 255 mμ for divinyl sulfide), as do the alkyl acyl sulfides or thiol acid esters.

Thioanisole is considerably different from thiophenol in its spectrum; there are maxima near 210 and 255 mμ and a submerged maximum about 275 mμ. Possibly the first two are benzene-like bands and the last a $n \rightarrow \pi^*$ transition, though other interpretations can be made. A considerable number of alkyl aryl sulfide spectra have been determined.[14]

Diphenyl sulfide has maxima at 250 and 274 mμ that are generally little affected by substituents, though *ortho*-substitution does have a small hypsochromic and hypochromic action attributed to steric hindrance.

c. Disulfides and Polysulfides.

Many alkyl disulfides show a broad maximum near 250 mμ ($\epsilon \sim$ 400),[15] but the 5- and 6-membered ring disulfides absorb at longer wavelengths (about 80 mμ longer for 5-membered and 40 mμ for 6-membered rings)

[13] H. H. Jaffe and M. Orchin, *op. cit.*, Chapter 17.

[14] E. A. Fehnel and M. Carmack, *J. Am. Chem. Soc.* **71**, 84, 2889 (1949).

[15] A. E. Gillam and E. S. Stern, "Electronic Absorption Spectroscopy," 2nd ed., pp. 69–71. Edward Arnold, London, 1957.

TABLE VI-2

ULTRAVIOLET ABSORPTION MAXIMA OF SOME SIMPLE SULFUR COMPOUNDS[a]

Sulfur compounds	Maxima, mμ (ϵ)
Alkyl sulfides and polysulfides	
Dimethyl sulfide	210(1020), 229s(140)
Diethyl sulfide	210(1800), 229s(140)
Dicyclohexyl disulfide	248(560)
Tetramethylene disulfide	295(300)
Trimethylene disulfide	334(160)
Diethyl tetrasulfide	290(2400)
Dicyclohexyl hexasulfide	325(6000)
Aralkyl and aryl sulfides[b]	
Phenyl ethyl sulfide	210(3.94), 256(3.90), 270s(3.40)
Phenyl *tert*-butyl sulfide	218(4.08), 266(3.19)
p-Tolyl methyl sulfide	256(4.00), 282s(3.01)
Diphenyl sulfide	250(4.08), 277(3.76)
p-Tolyl phenyl sulfide	230(3.92), 250(4.09), 274(3.76)
Sulfones[b]	
Phenyl methyl sulfone	217(3.83), 264(2.99)
p-Tolyl methyl sulfone	225(4.07), 267(2.73)
p-Anilino methyl sulfone	213(4.01), 269(4.30)
Diphenyl sulfone	235(4.24), 266f(3.33)
Sulfoxides[b]	
Diallyl	210–215(3.5)
Diphenyl	233(4.15), 265(3.32)
Phenyl methyl	238(3.61), 256(3.45)

[a] In ethanol.
[b] Log ϵ given in place of ϵ.

and with lower intensity. A few disulfides are transparent down to 210–220 mμ (e.g., dibenzyl, allyl, *tert*-butyl), a phenomenon attributed to the inability of the sulfur-adjacent group to provide electrons for octet expansion in the sulfur.

Tetrasulfides and hexasulfides absorb at longer wavelengths than the disulfides and with increased intensity (see Table IV-2).

d. Sulfones.

The dialkyl sulfones are generally transparent down to 210 mμ. In α,β-unsaturated sulfones resonance interaction accounts for a band near 210 mμ, and similar interactions are apparent in alkyl aryl and diaryl sulfones, though these have spectra that are generally benzenoid in character. The data are voluminous.

e. Sulfoxides.

Dialkyl sulfoxides have an absorption near 210 mμ ($\epsilon \sim 1500$) in nonpolar solvents; the hypsochromic shift in polar solvents indicates that the band is a $n \rightarrow \pi^*$ transition, probably of the sulfur lone-pair electrons. Obviously there is no analog of this band in the sulfones.

Methyl phenyl sulfoxide has a single strong band near 253 mμ, and diphenyl sulfone two bands at 226 and 267 mμ in cyclohexane. The latter may be considered strongly perturbed benzene-type bands.

f. Thiones.

These are analogs of carbonyl compounds, but the $n \rightarrow \pi^*$ band of the C=S linkage is at relatively long wavelengths, sometimes even in the visible.[16,17] See Table VI-3.

TABLE VI-3

ULTRAVIOLET–VISIBLE MAXIMA OF SOME C=S COMPOUNDS OR FUNCTIONS[a]

Compound or group	Conjugation band, mμ (ϵ)	$n \rightarrow \pi^*$ band, mμ (ϵ)
Cyclohexanethione		504 (c. 10)
Thiobenzophenone	315(15,100)	595 (177)
p,p'-Dimethoxythiobenzophenone	350(26,100)	570 (277)
p-Nitrothiobenzophenone	300(12,930), 310(12,450)	610 (133)
S—C=S (aliphatic)		460
O—C=S (aliphatic)		377
N—C=S (aliphatic)		365
O \C=S (aliphatic) O/		303
O \C=S (aliphatic) S/		357

[a] Compound data from reference 16, function data from reference 17.

2. NEAR INFRARED

The thiols have a very weak ($\epsilon \sim 0.1$) overtone of the S—H stretching mode at 1.97–1.98 μ.

[16] G. Oster, L. Citarel, and M. Goodman, *J. Am. Chem. Soc.* **84,** 704 (1962).
[17] M. J. Janssen, *Rec. trav. chim.* **79,** 464 (1960).

3. INFRARED

In thiols the S—H stretching fundamental is a weak band in the 2550–2600 cm^{-1} region; it does not shift to lower frequencies for hydrogen bonding as much as the hydroxyl band.

In most sulfur compounds there are C—S stretching bands that fall in the 600–800 cm^{-1} range and are usually weak. The C $=$ S stretching band of thiocarbonyl compounds is in the 1050–1250 cm^{-1}(s) region

TABLE VI-4

PRINCIPAL INFRARED BANDS FOR ORGANIC SULFUR COMPOUNDS[a]

Sulfur compounds	Frequency (cm^{-1})
C=S Stretching	
Thioureas	1130–1430
Thioamides	c. 1120
(RS)$_2$C=S	1050–1060
(RO)$_2$C=S	1210–1235
—C=C—C=S	1140–1155
Ar$_2$C=S	1215–1230
S=O Stretching	
Sulfoxides	1030–1070
Sulfones	1120–1160, 1300–1350
RSOOH	c. 1090
RSOOR'	1125–1135
(RO)$_2$SO (sulfite)	1170–1220
(RO)$_2$SO$_2$ (sulfate)	1150–1230, 1350–1440
Sulfonic acids and salts	1010–1080, 1150–1260
Sulfonyl chlorides	1160–1190, 1340–1375
Sulfonamides	1160–1180, 1330–1370
C—S Stretching	570–705
S—H Stretching	2550–2590
S—O Stretching	600–900
S—S Stretching	450–550

[a] The last four bands are weak, the others strong generally.

and varies in position with substituents and other factors in about the same way as the analogous carbonyl (see Table VI-4 and Fig. VI-I). The S—S stretching mode in disulfides is a weak band in the 450–550 cm^{-1} region.

Stretching frequencies of sulfur-oxygen bonds are all intense; the

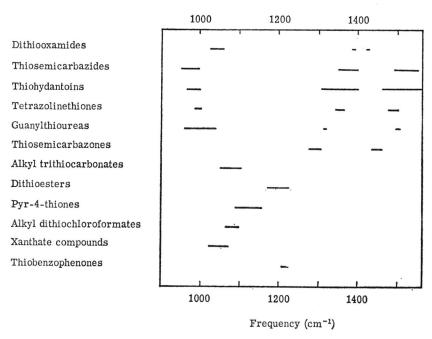

Frequency (cm⁻¹)

FIG. VI-1. Some C=S stretching bands in the infrared. [Adapted from C. N. R. Rao and R. Venkataraghavan, *Spectrochim. Acta* **18**, 541 (1962).]

S—O frequency is 700–900, the S=O 1040–1200 and the SO_2 an asymmetric and symmetric pair at 1310–1400 and 1120–1230 cm⁻¹. More precise correlations for specific functions can be made, and some of these are summarized in Table VI-4.

C. Phosphorus Compounds

1. ULTRAVIOLET–VISIBLE

Perhaps 1% of published ultraviolet spectra are of compounds containing phosphorus, since this element is a key one in substances of biological importance. There are few efforts to evaluate the effect of the phosphorus in these spectra, however; it is known that compounds in which the phosphorus retains a lone pair of electrons show an influence by the phosphorus on spectra while compounds without the lone pair often do not. Thus spectra of triphenylphosphines show a single major band that can be considered a highly perturbed benzene spectrum, but the aryl phosphonic and phosphinic acids and triphenylphosphine oxide

TABLE VI-5

PRINCIPAL INFRARED BANDS OF ORGANIC PHOSPHORUS COMPOUNDS

Phosphorus compounds	Frequency (cm^{-1})
Phosphorus-oxygen bonds	
P=O (free)	1250–1350
(hydrogen-bonded)	1150–1250 (s)
P—O—C	
(aromatic)	1190–1240 (s)
(aliphatic)	990–1050 (s)
P—O—Me	1180 (w)
P—O—Et	1150–1170 (w)
P—OH	2560–2700
Phosphorus-carbon bonds	
P—Me	1280–1320
P—Phenyl	1435–1450 (m)
P—H	2350–2440 (m)
P—O—P	910–970
P=S	600–840 (w)
P=N (phosphonitrilic chlorides)	1308–1315
P—N	c.715
P—Cl	440–580 (s)
P—F	810–885 (s)
Ionic phosphates	
Aryl substituents	1040–1090
Alkyl substituents	1080, 1150–1180 (s)

have typical benzene-like spectra, even to the extensive vibrational structure of the long wavelength band near 260 mμ.[18]

The cyclic phosphonitrilic chlorides, $(PNCl_2)_n$, are unlike carbocyclic analogs in that they have no maxima above 200 mμ. Extensive ultraviolet studies of phosphafluorinic acids,[19] phenoxyphosphonic acids[20] and cyclopolyphosphines[21] are among many papers on phosphorus compound spectra, none of them offering much interpretation of the data, however.

2. NEAR INFRARED

There is a moderately intense P—H overtone near 1.8 μ, but the P—H bond is a rather uncommon one.

[18] H. H. Jaffé and L. D. Freedman, J. Am. Chem. Soc. 74, 1069, 2930 (1952).
[19] L. D. Freedman and G. O. Doak, J. Org. Chem. 24, 638 (1959).
[20] J. N. Phillips, Australian J. Chem. 12, 199 (1959).
[21] W. Mahler and A. B. Burg, J. Am. Chem. Soc. 80, 6161 (1958).

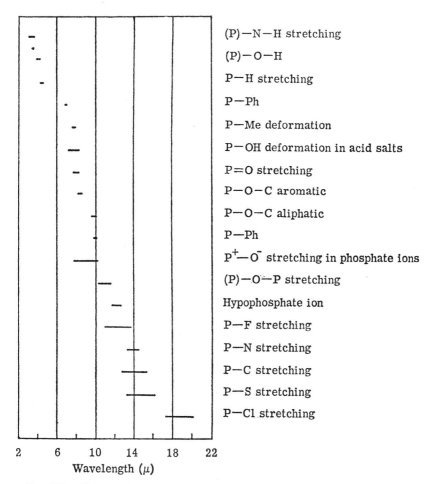

FIG. VI-2. Characteristic infrared bands of phosphorus compounds. Adapted from reference 12.

3. INFRARED

The P—H stretching fundamental is a sharp band in the 2350–2440 cm^{-1} region. The principal frequencies associated with the more common bonds of phosphorus to oxygen and other elements are summarized in Table VI-5 and Fig. VI-2.

D. Silicon Compounds

Since silica cells are commonly employed in spectrophotometry in the ultraviolet–visible and near infrared regions, it is plausible but incorrect

TABLE VI-6

INFRARED BANDS OF ORGANIC SILICON COMPOUNDS[a]

Silicon function	Frequency (cm^{-1})	Identification
Si—Me	1260	Symmetric Me deformation
	c. 800	Si—Me stretching
SiMe$_2$	1260	Symmetric Me deformation
	800–815	Si—Me stretching
SiMe$_3$	1250	Symmetric Me deformation
	840	
	755	Si—Me stretching
Si—H	2080–2280	Stretching
Si—O	1020–1090	Stretching
Si—S	19.5–22 μ[b]	Stretching
SiNSi	10.7 μ[b]	Asymmetric stretching
Si—F	10–12 μ[b]	Stretching
Si—Cl	16–24 μ[b]	Stretching

[a] All bands very strong.
[b] Wavelengths in microns.

to suppose them always fully transparent. The ordinary cells have absorption bands near 230 mμ in the ultraviolet and 2.7 μ in the near infrared, and special types are required when one of these must be absent.

Ultraviolet spectra of tetraphenylsilicon and several triphenylsilanes and derivatives have been compared to those of the corresponding carbon compounds. Few significant differences exist, and some resonance interaction through the silicon can explain these.

The major characterization of silicon compounds is in the infrared spectrum. The intensities of bands associated with silicon are generally about five times as great as for the corresponding carbon compounds.[22] The Si—H and Si—O stretching bands are near 2200 and 1020–1090 cm^{-1} respectively, and Si—C bonds are generally represented by an 800 cm^{-1} band. In silanes the Si—H stretching frequency decreases with increasing substitution, with aryl or unsaturated groups always giving absorption at higher frequencies than alkyl. Silicones contain 800 and 1260 cm^{-1} bands associated with Si—Me linkages as well as two Si—O bands in the 1000–1100 cm^{-1} range.

[22] A. L. Smith and N. C. Angelotti, *Spectrochim. Acta* **15**, 412 (1959); R. N. Kniseley, V. A. Fassel, and E. E. Conrad, *ibid.* **15**, 654 (1959).

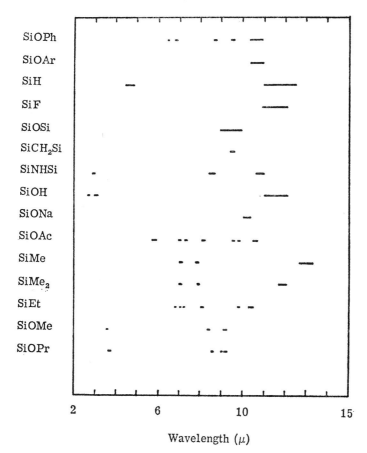

Wavelength (μ)

Fɪɢ. VI-3. Characteristic bands of some organosilicon functions. Adapted from reference 23.

Characteristic infrared absorptions for silicon compounds are summarized in Table VI-6 and in Fig. VI-3.[23]

E. Boron Compounds

Though ultraviolet spectra of a number of organic boron compound types have been published, the interpretative literature is slight. Decaborane, though not an organic compound, has some interest as a hydro-

[23] A. L. Smith, *Spectrochim. Acta* **16**, 87 (1960).

carbon analog; in cyclohexane it has a 272 mμ maximum.[24] Borazines have ultraviolet spectra resembling that of benzene as expected from the close structural resemblance; electron-releasing N-substituents have a bathochromic effect but B-substituents do not.[25]

Infrared spectra of boron compounds are summarized in Table VI-7. In trialkyl boron compounds the asymmetric B—C stretching frequency is in the 1116–1150 cm^{-1} range.[26] In boric acid esters an asymmetric B—O stretching is found at 1310–1350 cm^{-1}.[27]

TABLE VI-7

INFRARED ABSORPTION BANDS IN ORGANIC BORON COMPOUNDS[a]

Function	Frequency (cm^{-1})	Identification
B—H	1600–2220	Several bands
	2200	B—H stretching
	2500–2600[b]	Terminal B—H stretching in alkyl diboranes
BH$_2$	2500, 2600[b]	Asymmetric and symmetric stretching in alkyl diboranes
	1140–1205	In-plane deformation
	945–975	Out-of-plane deformation
B···H—B	1600–2000	Several bands
	1580–1610[b](s)	BH$_2$B bridges in alkyl diboranes
B—CH$_3$	1405–1460	Symmetric deformation
	1280–1320	Asymmetric deformation
B—Aryl	1430–1440(s)	
B—O	1310–1350(s)	Stretching
B—N	1330–1380	Stretching
B—Cl	890–910	Stretching

[a] All data are from reference 9, except as shown.

[b] W. J. Lehmann and I. Shapiro, *Spectrochim. Acta* **17**, 396 (1961).

[24] L. J. Kuhns, R. S. Braman, and J. E. Graham, *Anal. Chem.* **34**, 1700 (1962).

[25] D. W. Davis, *Trans. Faraday Soc.* **56**, 1713 (1960).

[26] W. J. Lehmann, C. O. Wilson, and I. Shapiro, *J. Chem. Phys.* **28**, 777 (1958).

[27] R. L. Werner and K. G. O'Brien, *Australian J. Chem.* **8**, 355 (1955).

VII

Inorganic Compounds

The correlation of structure with spectra for inorganic ions and co-ordination compounds in the ultraviolet–visible region has undergone a startling improvement in the last ten years. The development of the concept of charge-transfer as the source of many broad absorption bands and the widespread application of ligand field theory to predict absorption due to d-electron transitions in the transition metals are responsible for much of the improvement. Detailed assignments of bands to specific electron transitions are now commonly made.[1] The frequency unit of kilokayser (1 kilokayser = 1000 cm^{-1}) is preferred to wavelength units.

Infrared spectra of inorganic compounds tend to be simpler, or at least to exhibit fewer maxima at modest resolution, than for most organic compounds. A considerable number of far-infrared spectra of inorganic substances have also been published.

A. Ultraviolet–Visible

The cations of the alkali metals and alkaline earths are transparent down to 200 mμ. Aluminum, zinc, and cadmium ions do not absorb above 220 mμ. Tin, lead, mercury, and silver compounds may have appreciable absorption below 300 mμ, but this is largely the result of complex formation with suitable anions in the solution and thus depends on what ions are present.

The transition metals (elements with unfilled d-orbitals) are well known to exhibit characteristic ion colors in many instances; e.g., green for nickel, pink or blue for cobalt, blue-green for copper, etc. These visible absorption bands may be due to transitions between the levels of a single d^n configuration (d-d transitions) and are often relatively weak; there is also strong absorption in the ultraviolet in many of these ions as the result of charge-transfer mechanisms.

The coordination compounds of the transition metals constitute a large group of substances for which ligand field theory has been very

[1] C. K. Jørgensen, "Absorption Spectra and Chemical Bonding in Complexes," Pergamon Press, New York, 1962.

successful in predicting the weak band systems at relatively long wavelengths (sometimes extending into the near infrared). The theory has been most extensively applied to octahedral complexes of nickel, chromium, and cobalt, and Jørgensen[1] has tabulated a large portion of these data.

In the octahedral complexes the effect of replacing one ligand or coordinating group with another may be estimated from the spectrochemical series of ligands:

$$I^- < Br^- < Cl^- < F^- < ROH < H_2O < NH_3 < H_2NCH_2CH_2NH_2 < 1,10\text{-}$$
$$\text{phenanthroline} < NO_2^- < CN^-$$

The wavelength of maximum absorption for complex compounds of a given metal decreases from left to right in this series. Thus, the halogen complexes of a transition metal increase in λ_{max} with increasing atomic weight of the halogen. A spectrochemical series for central metal ions attached to the same ligand is as follows:

$$Mn(II) < Ni(II) < Co(II) < Fe(II) < V(II) < Fe(III) < Cr(III) < V(III)$$
$$< Co(III) < Mn(IV) < Mo(III) < Rh(III) < Ru(III) < Pd(IV) < Ir(III)$$
$$< Re(IV) < Pt(IV)$$

Presumably the wavelength of maximum absorption for a given group coordinated to these metals would decrease from left to right in this series also. (See Table VII-I.)

Colorimetric methods for the determination of metals often depend on

TABLE VII-1

ULTRAVIOLET–VISIBLE MAXIMA OF SOME TRANSITION METAL SPECIES[a]

Ion or compound	Maximum, mμ (ϵ)	Ion or compound	Maximum, mμ (ϵ)
$Ti(O_2)^{2+}$	429	$Cu(H_2O)_6^{2+}$	795
$Ti(H_2O)_6^{3+}$	492	$Cu(NH_3)_4^{2+}$	590
VO^{2+}	764 (16.5)	$Cuen_3^{2+}$	610
CrO_4^{2-}	275, 370	$Ru(H_2O)_6^{3+}$	224 (2300)
$CrCl_3$	440	$Pb(H_2O)_4^{2+}$	379 (86)
$Cr(H_2O)_6^{2+}$	710	WO_4^{2-}	199
$Mn(H_2O)_6^{3+}$	475	TcO_4^{2-}	247, 289
MnO_4^-	537	ReO_4^-	212, 232
MnO_4^{2-}	439, 605	$HgCl_4^{2-}$	230 (26,000)
$Co(NH_4)_2(SO_4)_2$	510	$PbCl_6^{2-}$	208 (24,000), 307 (9700)

[a] Where visible maxima exist, the principal one has been listed in most cases and ultraviolet bands omitted.

complex formation in solution with suitable ligands. The wavelength of the principal charge-transfer band in the resulting complex is longest for the most easily oxidized ligands, and in complexes of one ligand with a series of metal ions the longest wavelength absorption may be expected

TABLE VII-2

PRINCIPAL ULTRAVIOLET–VISIBLE MAXIMA OF SOME LANTHANIDE
AND ACTINIDE IONS[a]

Ion	Maxima, mμ (ϵ)
Nd(III)	354(5.05), 521.6(4.29), 575.5(6.93), 739.5(7.03), 794.0(9.96)
Pr(III)	444.2(10.31)
Sm(III)	401.5(3.31)
Eu(III)	394.2(3.01)
Gd(III)	272.8(3.35)
Dy(III)	350.4(2.52), 365.0(2.05), 908.0(2.38)
Ho(III)	287.0(3.31), 361.1(2.18), 450.8(4.00), 537.0(4.58), 690.4(3.05)
Er(III)	379.6(6.93), 523.5(3.20)
Tm(III)	683.0(2.56)
Yb(III)	908.0(1.97)
La, Lu	None from 200–1000 mμ
UO_2^{2+}	418(13.5)
Np(III)	233.5(2295), 267(1593), 522(44.5), 602(25.8), 661(30.5), 787.5(48.2)
Np(IV)	504(22.9), 590.5(16.1), 723(67.0), 743(43.0), 825(24.5), 964(185)
Np(V)	428(11.1), 617(23.7), 983
Np(VI)	476, 557
Pu(III)	603(35.4)
Pu(IV)	476
Pu(V)	775
Pu(VI)	833(300)
Am(III)	502.7(395), 814.0(50)
Am(V)	513.1(45.6), 715.1(59.3)
Am(VI)	666.0(75), 995.0(142)

[a] Lanthanide data (perchloric acid solutions) are from C. V. Banks and D. W. Klingman, *Anal. Chim. Acta* **15**, 356 (1956); actinide data from J. C. White, *in* "Analytical Chemistry," (C. E. Crouthamel, ed.), Volume 2. Pergamon Press, New York, 1961.

from the ion that is the strongest oxidizing agent. The notorious impermanence of many colored systems used in analysis is often the result of actual oxidation of ligand by ion.

Among cations the lanthanides and some of the actinides are unique in the sharpness of their absorption bands (Table VII-2), some of them in the near infrared. While these bands are extremely narrow and very

characteristic in wavelength for each element, they are not intense and
the visual appearance of the solutions is therefore not striking.

The far ultraviolet and ultraviolet spectra of anions contain many
charge-transfer band examples (see Table VII-3). In the ultraviolet
sulfate, perchlorate, borate, and silicate are transparent, and chloride
and the acid phosphates (HPO_4^- and HPO_4^{2-}) nearly so. Hydroxides
and cyanides are often contaminated with carbonate which absorbs
strongly below 230 mμ. Iodide, sulfite, thiocyanate, azide, thiosulfate,

TABLE VII-3

FAR ULTRAVIOLET AND ULTRAVIOLET MAXIMA OF SOME SIMPLE ANIONS[a]

Ion	Maxima, mμ (ϵ)	Ion	Maxima, mμ (ϵ)
Cl^-	181(c. 10,000)	OH^-	187(5000)
Br^-	190(12,000), 199.5(11,000)	SH^-	230(8000)
I^-	194(12,600), 226(12,600)	$S_2O_3^{2-}$	220(4000)
Br_3^-	270(11,500)	$S_2O_8^{2-}$	254(22)
I_3^-	287.5(40,000), 353(26,400)	NO_2^-	211(6000), 298s(8), 357(23)
ClO^-	190(strong), 292(330)	NO_3^-	193.6(8800), 302.5(7)
ClO_2^-	190(strong), 260(150)	$N_2O_2^-$	248(4000)
BrO^-	190(strong), 330(170)		

[a] All data are from S. F. Mason, *Quart. Revs.* **15**, 356 (1961); solutions are aqueous
or alcoholic.

bisulfide, and arsenate all absorb strongly below 260–280 mμ. Nitrates
have a weak band ($\epsilon \sim 7$) at 300 mμ and nitrites a stronger one at
355 mμ.[2]

Water is of course transparent in the ultraviolet–visible but shows
strong absorption below 185 mμ in the far ultraviolet and O—H stretching overtones in the near infrared that are relatively very strong; indeed,
probably no material can be more sensitively detected in the near infrared than water, and numerous methods for its determination are based
on this fact.

B. Infrared and Far Infrared

Both kinds of spectra for several hundred common inorganic compounds and ions have been determined by F. A. Miller and other workers.
Since such compounds have only a few atoms per molecule as compared
to common organic substances, the spectra are generally simpler than

[2] G. H. Beaven, E. A. Johnson, H. A. Willis, and R. G. J. Miller, "Molecular
Spectroscopy," Macmillan, New York, 1961.

those of organic compounds. Some representative results are shown in Figs. VII-1 and VII-2, and some far infrared bands are also listed in Table VII-4.

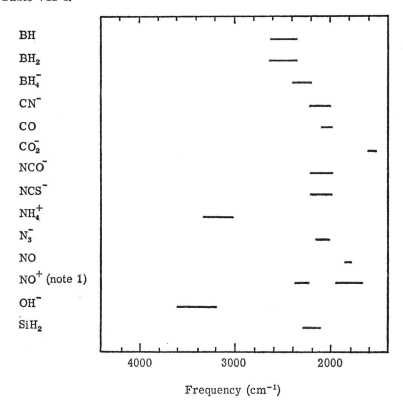

Frequency (cm^{-1})

FIG. VII-1. Infrared bands in the 1500–4000 cm^{-1} range for inorganic substances. Note 1: second band is for coordination compounds. Adapted from S. R. Yoganarasimhan and C. N. R. Rao, *Chemist-Analyst* **51**, 21 (1962).

Semiconductors such as germanium, indium antimonide, and gallium arsenide have characteristic absorption edges of interest for theoretical purposes in the infrared, and spectra of these as well as alkali metal halides and hydroxides have been studied in considerable detail by solid state chemists.[3] Many inorganic gases have surprisingly regular and simple infrared bands; ammonia, for example is routinely employed as a wavelength standard and the interpretation of the hydrogen chloride

[3] S. S. Mitra, *Solid State Phys.* **13**, 1 (1962).

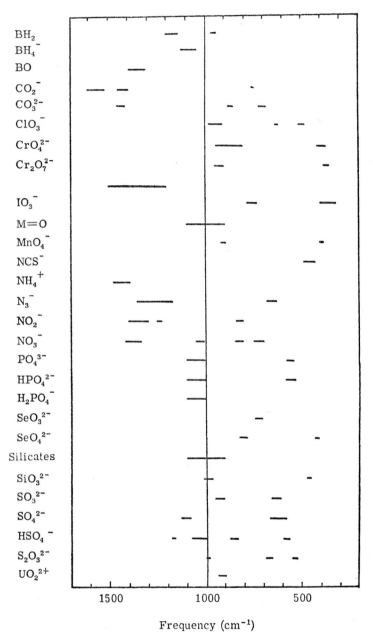

Frequency (cm⁻¹)

Fɪɢ. VII-2. Infrared bands in the 300–1500 cm⁻¹ range for inorganic substances. Adapted from S. R. Yoganarasimhan and C. N. R. Rao, *Chemist-Analyst*, **51,** 21 (1962).

TABLE VII-4

FAR INFRARED BANDS (300–700 cm⁻¹) OF SOME INORGANIC IONS[a]

Ion	Characteristic frequency (cm^{-1})
Carbonate	680–720
Bicarbonate	655–665(m), 690–710(s)
Thiocyanate	425–485 (doublet)
Phosphate	540–580
HPO_4^{2-}	530–580
Sulfite	615–660(s)
Sulfate	580–670
Bisulfate	570–600(s)
Thiosulfate	530–550, 660–690(s)
Chlorate	480–510(s), 615(s)
Perchlorate	620–630(s)
Bromate	360–370(s), 430–450(s)
Iodate	310–400(2 or 3 bands)
Chromate	370–420(w)
Dichromate	350–380(w)
Ferrocyanide	580–600
Ammonium, nitrite, nitrate, and cyanide	No absorption

[a] Data are from F. A. Miller, *Spectrochim. Acta* **16**, 162 (1960).

spectrum has been suggested as a suitable experiment for the undergraduate physical chemistry course. A compilation of over sixty spectra of common gases is available.[4]

Cotton[5] has discussed infrared spectra of many classes of coordination compounds, including many chelate compounds and also the sandwich compounds such as ferrocene and related organometallic systems.

A special note on the infrared absorption of water and carbon dioxide is necessary because these atmospheric components can cause difficulties in the interpretation of other spectra. Even in a double beam instrument that supposedly cancels out their effects by equal absorption in each beam sensitivity of measurements over regions where water and carbon dioxide absorb will be poor. The major bands of the carbon dioxide spectrum are at 2349.3, 720.5, and 667.3 cm⁻¹. Water vapor bands complete with much fine structure are encountered in several regions; principal absorption in liquid water is near 3400, 2050, 1640, and 700–800

[4] R. H. Pierson, A. N. Fletcher, and E. St. C. Gantz, *Anal. Chem.* **28**, 1218 (1956).

[5] F. A. Cotton, *in* "Modern Coordination Chemistry" (J. Lewis and R. G. Wilkins, eds.). Interscience, New York, 1960.

cm^{-1}. As is appropriate for the most common of all compounds, the spectrum of water has been studied in all possible regions of the electromagnetic spectrum, including the infrared as far as the borders of the microwave region.[6,7]

[6] E. K. Plyler and L. R. Blair, *J. Research Natl. Bur. Standards* **64C,** 55 (1960).
[7] N. G. Yaroslavskii and A. E. Stanevich, *Opt. Spectry. (USSR) (English Transl.)* **5,** 380 (1959).

VIII

Complex Materials

The selection of topics for this chapter has been largely dictated by the availability of useful reviews in the literature, and only a few of the industrially and biologically important classes of substances are considered. In general, infrared spectra are more likely to give useful data for truly complex materials than other spectra.

A. Polymers

The process of polymerization of an unsaturated monomer may often be tracked by ultraviolet spectrophotometry because the polymers are more nearly saturated and thus unlikely to absorb, or at least they should absorb at shorter wavelengths than the monomers.

Near infrared spectra of polymers have some utility as a means of following end groups such as the hydroxyl or as measures of aliphatic-to-aromatic ratios, but lack of solubility of most polymers in the few solvents that can be used in the near infrared is a handicap.[1]

Serious spectrophotometry for most polymers has been limited to the infrared region. Where a thin film can be prepared, the infrared spectrum is certainly very easy to get, and polystyrene films, for example, have been widely employed as standards for the calibration of wavelengths in infrared instruments.

Correlation charts and tables of absorption bands for common polymer types and functions have been compiled[2] (see Table VIII-1 for examples). A collection of 125 commercial polymer spectra[3] is useful for identifications, and a long classified bibliography in Elliott's review is also helpful.[4]

The infrared spectrum of polyethylene has been more closely studied than that of any other polymer since it is in theory the simplest of all.

---

[1] R. F. Goddu, *Advan. Anal. Chem. Instr.* **1**, 347 (1960).

[2] M. Tryon and E. Horowitz, *in* "Analytical Chemistry of Polymers, Part II," (G. M. Kline, ed.), Chapters VII–VIII. Interscience, New York, 1962.

[3] R. A. Nyquist, "Infrared Spectra of Plastics and Resins." Dow Chemical Co., Midland, Michigan, 1960.

[4] A. Elliott, *Advan. Spectry.* **1**, 214 (1959).

TABLE VIII-1

INFRARED SPECTRA OF SOME COMMON POLYMERIC STRUCTURES[a]

Polymer	Frequencies (cm^{-1})
Polyacrylates	1725–1750(vs), 1450(s), 1330(s), 1250(s), 1170(s), 850(m)
Polyadipates (crystal)	1725–1750(vs), 1470(s), 1420(s), 1275, 1175, 900
Polyacids	2700–3300(broad), 1700(s)
Polyketones	1700(vs)
Polyurethanes	3275–3400(s), 1700(vs), 1530(s)
Polyamides, RNHCOR'	3300(broad), 1675(vs), 1540(s)
Polyamides, RR'NCOR''	1675(vs)
Polydienes	1675(m)[b]

[a] All data are from reference 2.
[b] And other bands depending on substitution at the double bonds.

Data as far as 600 μ have been reported.[5] The degree of crystallinity in polyethylene can be estimated from the splitting of the 728 cm^{-1} rocking mode.

Orientation in polymer fibers and chains is perhaps more important than the particular functional group in the repeating unit in many polymers. A polarized infrared source permits measurements of absorption differences as a function of orientation.[4]

B. Sugars, Polysaccharides, Cellulose, and Wood

All of these materials are based on a tetrahydropyran nucleus in progressively increasing degrees of aggregation from the simple sugars through the polysaccharides to the typically polymeric celluloses. The predominant functions are OH and CH, and since there may be many of these per molecule differing only slightly in absorption maxima, the OH and CH regions of the infrared are likely to exhibit rather poorly defined bands. It has been suggested that really distinctive results for carbohydrates might be expected only in the 730–960 cm^{-1} range where overall molecular vibrations occur, and a table of sugar pyranose ring bands in this region has been assembled.[6a] More complete spectra of

[5] T. K. McCubbin, Jr., and W. M. Sinton, *J. Opt. Soc. Am.* **40**, 537 (1950).

[6a] W. B. Neely, *Advan. Carbohydrate Chem.* **12**, 13 (1957); H. Spedding, *in* "Methods in Carbohydrate Chemistry," (R. L. Whistler and M. L. Wolfrom, eds.). Academic Press, New York, 1962.

numerous aldopyranosides and derivatives may be found in the works of Tipson and Isbell.[6b]

Cellulose and wood spectra have been discussed by Marchessault.[7]

C. Steroids

Since the principal functional groups encountered in steroids are hydroxyl and carbonyl functions and multiple bonds, the application of infrared group frequencies for characterization is obvious. An atlas of steroid spectra is available.[8] Fieser[9] has tabulated functional group frequencies for various steroid types.

The far ultraviolet region may be used to detect isolated double bonds in steroids (see Chapter II), and the weak carbonyl maximum near 280 mμ in the ultraviolet is also valuable for characterization because α-halogens or α-acetoxy groups increase its intensity and wavelength differences for axial and equatorial conformations also exist.[9]

Steroids with conjugated systems have been studied in much detail in the ultraviolet region.[10] A conjugated double bond system is characterized by a strong maximum in the 220–250 mμ ($\epsilon \sim$ 14,000–28,000) region if the double bonds are in different rings (heteroannular) but in the 260–285 mμ ($\epsilon \sim$ 5000–15,000) region for double bonds in one ring (homoannular). An additional conjugated double bond yields a 30 mμ bathochromic shift, and alkyl substituents and exocyclic double bonds are each worth a bathochromic increment of about 5 mμ, following Woodward's rules in this respect. Most steroid spectra are determined in ethanol, but solvent corrections for other solutions can be applied.

The α,β-unsaturated ketones in the steroid series have been similarly analyzed. There is the weak carbonyl maximum near 290 mμ but a strong band for the conjugated system at 215 mμ is of greater importance. This band undergoes a 30 mμ bathochromic shift for an additional conjugated double bond, and 15, 23, and 35 mμ increments for α-chloro, bromo, or hydroxyl groups respectively. A "homodiene" shift of 39 mμ is also used

[6b] R. S. Tipson and H. S. Isbell, *J. Research Natl. Bur. Standards* **65A,** 249 (1961); **64A,** 239, 405 (1960).

[7] R. H. Marchessault, *Pure Appl. Chem.* **5,** 107 (1962).

[8] K. Dobriner, E. Katzenellenbogen, and R. N. Jones, "Infrared Absorption Spectra of Steroids, An Atlas," Volume I. Interscience, New York, 1953; Roberts, Gallegher, and Jones, Vol. II, 1958.

[9] L. F. Fieser and M. Fieser, "Steroids." Reinhold, New York, 1959.

[10] L. Dorfman, *Chem. Revs.* **53,** 47 (1953).

when necessary, and various other substituent correlations can also be made.

The identification of steroids through computer analysis of their ultraviolet absorption at several wavelengths appears likely to enjoy considerable success.[11]

D. Amino Acids, Polypeptides, and Proteins

Since polypeptides and proteins are successively built up from amino acids in various combinations by amide formation, it is evident that peptides and proteins can be characterized in the infrared by amide bonds (see Chapter IV). Elaborate reviews of the spectra of all these classes of compounds in both ultraviolet and infrared regions are available.[12-15]

1. Far Ultraviolet

Polypeptides have been found to show a rough additivity of light absorption by the individual peptide links, with the amide maximum near 185 mμ doubling for each such linkage added to the molecule.

2. Ultraviolet–Visible

Of the approximately twenty amino acids commonly found in proteins only tyrosine, tryptophan, phenylalanine, and cystine have any characteristic absorption in the 220–300 mμ region of the spectrum,[12] the first three of these by virtue of an aromatic ring and the last through its disulfide linkage. Tryptophan and phenylalanine do not vary much in spectrum as the pH changes and the acids ionize, but tyrosine contains a phenolic hydroxyl that accounts for marked changes in spectra between acidic and basic solutions.

In proteins the aromatic amino acid bands are largely obscured and a general broadening of the absorption curve relative to the constituent amino acids occurs. There is also a bathochromic shift of the absorption as compared to mixtures of amino acids. These effects result from hydro-

[11] E. R. Garrett, J. L. Johnson, and C. D. Alway, *Anal. Chem.* **34,** 1472 (1962).

[12] J. P. Greenstein and M. Winitz, "Chemistry of the Amino Acids," Volume 2, Chapter 17. Wiley, New York, 1961.

[13] G. H. Beaven in *Advan. Spectry.* **2,** 331 (1961).

[14] P. Doty and E. P. Geiduschek, *in* "The Proteins," (H. Neurath and K. Bailey, eds.), Chapter 5. Academic Press, New York, 1953.

[15] C. Tanford, "Physical Chemistry of Macromolecules." McGraw-Hill, New York, 1962.

gen bonding and other structural factors that make the environment of the amino acid functional groups very different in a protein from that in the acids.[13]

Much use has been made of differential spectrophotometry in ultraviolet work with proteins. In this technique the spectrum of the protein after treatment with acid, base or any other reagent of interest is measured against the initial protein solution in place of the usual solvent blank. Only the differences produced by the treatment of the protein will then show in the spectrum.

3. INFRARED

Extensive investigations of infrared bands in all the ordinary amino acids have been summarized in considerable detail by Greenstein and Winitz,[12] who tabulate principal bands in the 2–8 μ region for over fifty of them (see Table VIII-2). In the solid state the amino acids are pre-

TABLE VIII-2

PRINCIPAL INFRARED BANDS (2–8 μ) OF SOME COMMON AMINO ACIDS[a]

Compound	Principal bands (μ)
Glycine	3.17, 3.50, 3.90, 4.75, 6.23, 6.60, 6.93, 7.08, 7.50
Alanine	3.30, 3.73, 3.90, 4.77, 6.15, 6.28, 6.65, 6.88, 7.08, 7.35, 7.65
Valine	3.38, 3.75, 4.79, 6.35, 6.67, 7.02, 7.18, 7.40, 7.53, 7.87
Leucine	3.39, 3.75, 3.93, 4.75, 6.34, 6.62, 6.95, 7.10, 7.35, 7.61, 7.72
Isoleucine	3.44, 4.77, 6.32, 6.60, 6.84, 7.06, 7.16, 7.40, 7.52, 7.64
Tyrosine	3.25, 3.75, 4.85, 6.22, 6.30, 6.62, 7.06, 7.34, 7.50, 8.03
Tryptophan	2.96, 3.34, 4.00, 4.85, 5.98, 6.28, 7.07, 7.36
Serine	2.95, 3.35, 3.80, 4.92, 6.25, 6.81, 7.08, 7.46, 7.67
Threonine	3.20, 3.40, 4.90, 6.14, 6.77, 6.87, 7.05, 7.43, 7.60, 8.00
Proline	3.33, 6.18, 6.43, 6.92, 7.29, 7.57, 7.75, 7.98
Aspartic acid	3.43, 4.85, 5.25, 5.93, 6.06, 6.25, 6.65, 7.03, 7.42, 7.57, 8.01
Histidine	3.40, 6.11, 6.27, 6.57, 6.84, 7.06, 7.43, 7.60, 7.84, 7.98
Cystine	3.47, 4.80, 6.16, 6.33, 6.74, 7.10, 7.24, 7.47, 7.72

[a] All data from Greenstein and Winitz.[12]

dominantly in the dipolar or zwitter ion form, and thus possess a 6.3 μ band associated with the carboxylate group and —NH$_3^+$ bands near

4.7 μ (weak). Many amino acids are much alike in the 6.3–7.5 μ region, showing absorption at 6.3, 6.6, 6.9, 7.1, 7.3 and 7.5 μ. The 6.3 and 7.1 μ bands are respectively asymmetric and symmetric stretching of the ionized carboxyl, the 6.6 μ band is a N—H deformation mode, and the others are associated with methyl and methylene groups. According to Bellamy NH stretching in amino acids is observed as a single band in the 3030–3130 cm^{-1} range.

Diastereomeric amino acids may generally be easily distinguished in the infrared. The L and D isomers of the amino acids, however, have identical infrared absorption, though the DL-acids as crystallized from solution often are somewhat different.

In the spectra of amino acid hydrochlorides there is an unionized carboxylic acid function represented by bands at 5.76, 5.80, 6.00, and 6.07 μ, and the carboxylate bands at 6.3 and 7.1 μ in the free amino acids are absent.

In the near infrared proteins have some combination bands, notably one at about 4600 cm^{-1}, that are interesting because they can be measured on aqueous solutions.

Infrared spectra of proteins are usually determined on solid state samples and the complexities expected from the presence of multiple amide linkages and hydrogen bonding are fully evident. Hydrogen bonded N—H groups account for bands near 3300 cm^{-1} and amide linkages for two main bands in the 1500–1600 cm^{-1} region. Since the extended chains or helices of protein structures may absorb differently when the light vector runs the length of the structure rather than across it, measurements with polarized infrared on oriented samples permit useful deductions concerning the placement of carbonyl and other functions.

E. Purines, Pyrimidines, Nucleic Acids, and Related Materials

The ultraviolet spectrum of pyrimidine (see also Chapter V) is characterized by a $\pi \rightarrow \pi^*$ band at 245 mμ and a weaker $n \rightarrow \pi^*$ absorption at 300 mμ; the latter band may be reduced to a shoulder by many of the substituents occurring in nucleotide bases. Purines generally have two $\pi \rightarrow \pi^*$ bands in the 230–280 mμ region (see Table VIII-3 for illustrative spectra).

The purines and pyrimidines that are of biological significance usually have acidic hydroxyl or basic amino groups (or both) as substituents on the basic heterocyclic structure. Substantial variations in spectra as

TABLE VIII-3

ULTRAVIOLET MAXIMA OF SOME PURINES AND PYRIMIDINES
OF BIOCHEMICAL IMPORTANCE[a]

Compound	Maxima, mμ (log ϵ)		
	pH 1	Neutral	pH 12
Purine	258(3.8)	261(3.9)	270(3.90)
Xanthine	265(3.90)	268(3.99)	241(3.92)[b] 278(3.95)
Hypoxanthine	250(3.96)	250(4.02)	
Adenine	260(4.13)	260(4.1)	267(4.08)
Guanine	250(4.00)	246(4.14) 275(4.02)	
Thymine	265(3.88)	264(3.89)	291(3.74)
Uric acid	231(3.9) 285(4.07)	238(4.00) 293(4.10)	296(4.1)[c]
Caffeine		278(4.04)	
Uracil	260(3.91)	259(3.91)	284(3.79)
1-Methyluracil		268(3.99)	265(3.85)
2-Thiouracil	278(4.09)	274(4.03)	230(4.08) 259(4.04)
Cytosine	275(4.00)	267(3.79)	272(3.75)
Barbituric acid	205(4.02) 258(2.78)	255(4.39)	258(4.29)
Alloxan		243(3.40)	

[a] All data from Volumes I and II of "Organic Electronic Spectral Data" [M. J. Kamlet (Vol. I) and H. E. Ungnade (Vol. II) eds.]. Interscience, New York, 1960.
[b] pH 10.5.
[c] pH 13.

functions of pH can be expected from these compounds. Spectra at four or five pH values from one to thirteen thus offer excellent characterization of adenine, hypoxanthine, guanine, uracil, thymine, cytosine, and related compounds.[16] Determinations of ionization constants through these changes of spectra may also be performed (see Chapter III). New purines and pyrimidines are frequently characterized by their absorption

[16] G. H. Beaven, E. R. Holiday, and E. A. Johnson, *in* "The Nucleic Acids" (E. Chargaff and J. N. Davidson, eds.), Chapter 14. Academic Press, New York, 1955.

maxima at pH 1 and pH 13 in the ultraviolet; Robins,[17] for example, has published a number of papers each containing a hundred or more such spectra.

Nucleosides resemble the purine and pyrimidine bases that are parts of their structure to a considerable degree, since the prime structural difference is the presence of a sugar moiety that usually has little distinctive absorption of its own in the ultraviolet. Tables of maxima and minima in the ultraviolet spectra of adenosine, inosine, guanosine, xanthosine, cytidine, thymidine, uridine, and other nucleosides are available.[16] The dissociation of a 2-hydroxy group in many pyrimidines and of an iminazole NH in purines are responsible for spectra changes in the pH 12–13 region for these compounds that cannot occur in some nucleosides where these functions are blocked.

Most nucleotides are very little different from the corresponding nucleosides in their ultraviolet spectra.

The nucleic acids are noteworthy in that their absorptivities in the ultraviolet are about 35% lower than expected from the summation of component purine and pyrimidine nucleotides. This hypochromism may result from the intramolecular organization of these high molecular weight substances into the well known double helix with many hydrogen bonds; in such a structure the individual chromophores could not be independent of each other.

Nucleoproteins exhibit ultraviolet bands for both the nucleic acid and protein portions of their structures; since the proportions as well as the nature of these components may vary widely, it can only be said that the nucleic acid absorption is generally dominant and the protein contribution from its aromatic amino acid residues is likely to appear only as a shoulder near 280 mμ.

Studies of nucleic acids and nucleoproteins in the ultraviolet with polarized radiation and oriented samples have helped to confirm X-ray evidence for the structures of these significant materials. The dichroism is calculated from the difference in absorptivities when the electric vector is parallel rather than perpendicular to the fibrous structure.

Infrared spectra of all the materials mentioned in this section have been determined but offer unusual problems of interpretation. In the biologically significant purines and pyrimidines there are hydroxyl and amino substituents on the nucleus that allow amino-imino and keto-enol tautomeric structures. Several absorptions in the single and double

[17] R. K. Robins, *et al.*, *J. Med. Pharm. Chem.* **5**, 1085 (1962), and earlier papers.

bond regions are certain. Bellamy does not make any precise correlations of structure with spectra for these substances.

It is of interest that the vastly more complex infrared spectra of bacteria and microorganisms generally have also been subjected to attempted analysis.[18] It appears possible to differentiate related strains in many instances but the differences are quite small and the work is difficult and uncertain.

F. Ultraviolet–Visible Spectra of Porphins, Chlorins, and Tetrahydroporphins

These 18-membered ring systems consisting of four pyrrole rings bridged by methylenes are of exceptional interest because of their occurrence as iron and magnesium chelate compounds in chlorophyll, hemoglobin, and other biologically important materials in which color is an obvious factor in the biological function.

In general the visible spectra of porphins consists of four moderately intense bands, numbered I–IV in order of decreasing wavelength, and a fifth intense band ($\epsilon \sim 100,000$), called the Soret band, near 400 mμ. Bands I and III are particularly sensitive to the nature, position and number of substituents, and empirical rules for these effects have been established.[19]

Platt has explained the spectra of these compounds with the use of a molecular orbital free electron model. Bands I–IV are forbidden transitions and the Soret band obviously an allowed transition. The degeneracy of the forbidden transition is broken in porphin itself, bands I and III being the band origins of the two components. Bands II and IV are vibrational absorptions. In the spectra of porphin metal chelates the degeneracy of the forbidden band is not split, and the longer wavelength band (I) is the band origin of both components.

Substitution in porphin will generally increase the intensity of the absorption of light as the forbidden transitions thus acquire an allowed character. Reduction to chlorins or tetrahydroporphins has a similar effect.

G. Visible Spectra of Dyes

With the exception of a few molecules of unusual theoretical significance (Malachite Green, methylene blue, phenol blue) there has been

[18] K. P. Norris, *Advan. Spectry.* **2**, 293 (1961).

[19] S. F. Mason, *Quart. Revs. (London)* **15**, 334 (1961).

little interest in the spectra of dyestuffs in recent years. Considering the fundamental role these materials played in the development of theories of color and their industrial importance, this must seem surprising. For most dyestuffs wavelengths of maximum absorption are readily available data but molar absorptivities are not. No doubt it may be safely assumed that the molar absorptivity of a dye at a maximum in the visible must be extremely high. Absorption bands for a few typical dyes are listed in Table VIII-4.

TABLE VIII-4

VISIBLE ABSORPTION MAXIMA OF SOME COMMON DYESTUFFS[a]

Substance	Maxima (mμ)	Substance	Maxima (mμ)
Rhodamine B	517, 557	Fluorescein	460, 494
Malachite green	617	Methylene blue	609, 668
Crystal violet	541, 591	Rosaniline	488, 547
Cyanine	550, 586	Kryptocyanine	655, 712 (EtOH)
Congo red	497	Orange I	488
Indigocarmine	617	Phenosafranine	494, 525
Trypan blue	585	Indigo	605 (CHCl$_3$)
Chrysoidin	461	Night green	635
Ponceau 2R	499, 538	Pyronin G	510, 551
Eosin FF	484, 518	Tartrazine O	432

[a] In aqueous solution except as shown.

Cyanine dyes have received more careful attention from spectroscopists than other dyes.[20] In these substances there is an alternating double bond system connecting two heterocyclic base nuclei, one carrying a positive charge. An increase in the number of conjugated double bonds increases the wavelength of maximum absorption in an approximately linear proportion (compare with polyenes, Chapter II). The precise value of the wavelength for a given number of double bonds depends on the relative basicities of the heterocyclic nuclei, and numerical factors for these have been worked out that permit a rather close prediction of the principal absorption band for a new compound in this series.

[20] L. G. S. Brooker, *Rev. Mod. Phys.* **14**, 275 (1942).

Author Index

Numbers in parentheses are footnote numbers. They are inserted to indicate the reference when an author's work is cited but his name is not mentioned on the page.

A

Adolph, H. G., 109
Adelfang, J. L., 92
Albert, A., 120
Alberti, C., 123
Allinger, J., 72
Allinger, N. L., 72
Altshuller, A., 106
Alway, C. D., 150
Amiel, Y., 10
Andrussow, K., 84
Ang, K.-P., 84
Angelotti, N. C., 136
Applewhite, T. H., 1
Armitage, J. B., 10, 47
Arnaud, P., 73
Arriel, Y., 44

B

Barrow, G. M., 7
Bauer, W. H., 18, 31, 35(8)
Bauman, R. P., 19, 128
Beaven, G. H., 36, 48(22), 54(22), 127, 142, 150, 151(13), 153, 154(16)
Beck, M. T., 83
Beer, M., 47
Bell, R. P., 82
Bellamy, L. J., 17, 33, 36, 54(10), 75, 77, 117
Ben-Efraim, D. A., 42, 44(33)
Bentley, F. F., 18, 32(11), 33(11), 34, 41(11), 55(11), 56(11), 61(11), 77
Bergmann, E. D., 62
Beringer, F. M., 126
Berman, E., 27
Bertelli, D. J., 62
Berthier, G., 77
Biggs, A. I., 82, 84

Bladon, P., 38
Blair, L. R., 146
Blanc, J., 49
Bohlmann, F., 42, 43, 44(34), 49
Booker, H., 42
Boulton, A. J., 123
Bower, V. E., 82
Bowman, R. E., 69
Boyd, R. H., 84
Braman, R. S., 138
Braude, E. A., 9, 79
Brecher, C., 49
Brewster, R. Q., 93
Brocklehurst, P., 65
Brooker, L. G. S., 156
Brown, T. L., 77
Bruck, P., 36
Bryson, A., 82
Buc, H., 69
Bunce, S. L., 18, 31, 35(8)
Bunnenberg, F., 106
Burawoy, A., 65
Burg, A. B., 134

C

Cairns, T. L., 110
Calvin, M., 43
Cannon, C. G., 60
Carlson, G. L., 18
Carmack, M., 129
Cederholm, B. J., 67
Chan, S., 94
Citarel, L., 131
Clar, E., 57
Coburn, W. C., Jr., 57
Coffin, E., 95
Coffman, D. D., 110
Coggeshall, N. D., 88, 90(68)

157

Subject Index

Entries marked with asterisk are in tables or charts on indicated page.

A

Absorptivity, molar, 4
　range, 4, 9–10, 12
Accuracy, 19–21
Acenaphthene, 57*
Acenaphthylene, 61
Acepleiadylene, 62*
Acetaldehyde, 71, 74*, 77
　chloro-, 74*
　dichloro-, 74*
　trichloro-, 74*
Acetaldoxime, 79
Acetamide, 101
Acetanilides, 102*
Acetate(s), 85*, 88*, 89
　phenolic, 85*, 88*
Acetic acid, 86
Acetone, 70, 71*, 77
Acetonitrile, 109
Acetophenones, 73–74, 77
　o-substituted, 74
Acetyl chloride, 87
Acetylene, 46
　halo derivatives, 48
Acetylenedinitrile, 107*, 109
Acetylenes, 46–48
　disubstituted, 47
　hot bands in, 47
　monosubstituted, 47, 48*
Acid anhydrides, 86–89, 88*, 127*
Acid bromides, 87
Acid chlorides, 86–89
Acid halides, 86–89, 127*
Acid iodides, 87
Acids, carboxylic, 80–86, 127*
　aliphatic, 80, 85
　aromatic, 86
　carbonyl band in, 85–86, 127*
　dibasic, ionization of, 84
　dimers, 84–85

α-halo-, 85
o-hydroxyaromatic, 86
ionization of, 80–84
salts, 86
α,β-unsaturated, 85*
Acridines, 120
Acrylates, 88*
Actinides, 141–142
Adenine, 153*
Alanine, 151*
Alcohols, 67–70, 85*, 91*
　aliphatic, 67
　aromatic, 68
　cyclic, 70
　higher, 70
　hindered, 70
　primary, 68
　secondary, 70
　tertiary, 70
　unsaturated, 68
Aldehyde C-H frequency, 76
Aldehydes, 70–77
　alkyl, 71
　aryl, 73
　combination bands, 75
　hydrogen stretching in, 31*
　β-hydroxy-α,β-unsaturated, 73*
　α,β-unsaturated, 72, 73*, 74*
Aldopyranosides, 149
Alkanes, 32*, 34
1-Alkenes, 37*, 38, 41
　2-alkyl-, 38
2-Alkenes, 38
Alkyl halides, 125–128
N-Alkylpyridinium compounds, 115
Alkynes, 46–48
Allene, 49
Allenes, 49
Alloxan, 153
Amides, 101–103